LIFE'S ENGINES
||||||||

LIFE'S ENGINES

||||||||

How Microbes Made Earth Habitable

Paul G. Falkowski

With a new preface by the author

PRINCETON UNIVERSITY PRESS

PRINCETON AND OXFORD

Published by Princeton University Press, 41 William Street, Princeton,
New Jersey 08540
In the United Kingdom: Princeton University Press,
99 Banbury Road, Oxford OX2 6JX
press.princeton.edu

First Princeton Science Library paperback edition, 2023
Paperback ISBN 9780691247687
E-book ISBN 9780691247694

Library of Congress Control Number: 2022949622

Cover design by Michael Boland for TheBolandDesignCo.com
Cover image: Artwork showing a variety of microbes. This artwork depicts viruses,
bacteria, fungi, algae and protozoa (organisms are not drawn to scale
relative to each other). Carlyn Iverson / Science Source

British Library Cataloging-in-Publication Data is available
This book has been composed in Goudy Old Style

To my parents, Ed and Helen, my wife and friend,
Sari Ruskin, and our daughters, Sasha and Mirit

||||||||

Contents

॥॥॥॥॥॥

Contents

Preface to the Princeton Science Library Edition

ⅠⅠⅠⅠⅠⅠ

*L*ife's *Engines* is essentially a mystery book. The mystery is how microbes came to make Earth habitable for us and virtually all other animals. The "detectives" are the scientists who discovered microbes and who have worked for the past 350-odd years to understand them. While the story is about science and is semi-autobiographical, fundamentally it is about solving the mystery of how the world in which we live came to be.

Two main questions at the heart of this book are "How did life originate?" and "Are we alone?" Although these two questions may appear to be trite, they remain unanswered. Microbes, as you will see, are at the core of both.

Although Darwin suggested that life could have arisen in a "small warm pond," its actual origins have been debated for decades. A century ago, in 1924, a Soviet biochemist and plant biologist, Alexander Oparin, wrote, "Thus, life is a special, very complex form of motion of matter, but this form did not always exist, and it is not separated from inorganic nature by an impassable abyss; rather, it arose from inorganic nature as a new property in the process of evolution of the world. We must study the history of this evolution if we want to solve the problem of the origin of life." What neither Darwin nor Oparin knew was whether life arose because of light being a driving force, or because of the existence of deep-sea hydrothermal vents, which were discovered only in 1977. Today, light provides virtually all the energy for life on Earth. Indeed, every chemical bond in our bodies, and in every other animal, is stored solar energy. However, deep-sea hydrothermal vents, which are essentially undersea volcanos, can also supply energy and are hypothesized to have been where life originated. In the end, we are still discussing and debating this issue. But why is the answer so elusive?

Over the past three centuries, science has mostly been guided by the concept of "reductionism," that is, an attempt to understand how things work by taking them apart. Reductionism has influenced how scientists have thought about the origins of life. Imagine taking home a bag of groceries, stomping on it until everything was mushed up, and then asking someone to tell you what you had bought. For a long time, that is how biologists and chemists analyzed living cells; we ground them up and separated the parts, analyzed their components, and tried to describe what each part does. We have learned a lot from this approach, especially over the past couple of decades, when it became possible to quickly analyze the constituents of the cells, such as their genes and proteins. In effect, we created a high-tech "parts department" for life. But, the bridge from the "parts" to the emergence of even the simplest of the nanomachinery that gives life to cells remains obscure.

The philosophical opposite of reductionism is emergence. Life is an emergent property. It uses the "parts" to create self-assembling nanomachines. In 1999, a French microbiologist, Patrick Forterre, suggested that all life arose from a single microbe, which he dubbed the Last Universal Common Ancestor (LUCA for short). It is a hypothetical organism that no longer exists, but we know, for sure, it was a microbe. How LUCA, the first self-assembling nanomachine, arose remains unclear, but it contained instructions for their formation that were passed via evolution to myriad microbes and ultimately to all plants and animals. We all share the same machinery that makes our bodies and gives us the power to make them again, and again, and again, ad infinitum. The bodies may undergo extinctions over geological time, but the processes remain the same. In effect, a single event in Earth's history became the foundation of all life as we know it.

During the time this book was written, about a decade ago—and now even more so—we have become increasingly aware that microbes rule virtually all aspects of our planet, as well as our bodies. Like many realizations in science, this largely came about because of advances in technology. We now have machines that can sequence an entire microbial genome in a few hours. What does that mean? The average microbial genome contains about 5 million base pairs, that is, 10 million small parts in a strict sequence, and that information dictates the construction of all known biological nanomachines. With currently available technology, humans can build this set of instructions for a few hundred dollars. Simultaneously, computer technology has become so advanced that we can "read" the genome sequences very rapidly and determine what the microbes do to create life, how many microbes there are, and their associations with each other. Similarly, advances in microscope technologies— far more advanced than what Leeuwenhoek, the first person to see microbes under a microscope (in 1676), could ever have imagined—now allow us to see microbes in the environment and to "tag" specific organisms with dyes.

Together, technologies have revealed that our planet harbors microbes virtually everywhere we look, from many meters down in sediments of the ocean, to ancient ices at the poles of our planet, to deep in soils, on roots of every plant, on the surface of every rock, even in clouds. And, as we now know, they not only rule our planet; they rule our bodies. We are conceived without microbes, but as we pass through the vaginal canal, they enter our bodies. As babies, we acquire still more microbes from our mothers, from dirt we ingest, and from other people. As we grow, each of us acquires a set of microbes that are unique to us. It is estimated that we have 40 trillion microbial cells in our guts—a figure equal to the total number of all the cells in our bodies—and that our microbes account for about 200 to 300 grams or so of our adult weight. That small biomass is critical to digesting the food we eat. Indeed, microbes are in the guts of every animal. Gut microbiomes are microcosms of the planetary microbiome. Many questions are arising about how these microscopic organisms work collaboratively both within the guts of animals and on a planetary scale. Indeed, the total biomass of microbes on Earth is less than 15% of the biomass of all life.

Also, over the past decade, many structures of the nanomachines within microbes have been discovered. These structures, which are maintained in the Protein Data Bank at Rutgers University, offer, in effect, an ultra-high-end view of the "parts department," where we now can literally visualize the microscopic biological pistons, springs, and gears, but still cannot reconstruct the basic engines of life. Over the coming decades, we undoubtedly will discover more about how the nanomachines within microbes work. But it may take yet more time to reconstruct life in a test tube and to actualize Oparin's dream.

One of the fundamental messages I got from writing *Life's Engines* is that all organisms exchange gases with their environment. That is obvious on a local, personal level, but not so obvious on a planetary level. Microbes made the first oxygen on Earth. The late Carl Sagan suggested that the discovery of significant oxygen on another planet would be a sign of life. He wrote, "The origin and evolution of life are connected in the most intimate way with the origin and evolution of the stars." In the years since Sagan's death in 1996, astronomers have discovered more than 5000 planets orbiting stars within our own galaxy, the Milky Way. At the time of writing this preface, the first images from the James Webb telescope have been displayed in the popular press. Undoubtedly, many thousands more extrasolar planets will be discovered in the next decades. The search for life far from our planet almost certainly will show that we are not alone.

Science should be fun. I hope that *Life's Engines* engages you, the reader, and helps you appreciate the world you cannot see with your naked eyes, the microbial life on this planet, or the outcome of its presence on planets far out in space.

Acknowledgments

||||||||

I had been thinking about this book for some time and wrote it in fits and starts over a period of almost two years. The core ideas were developed as an outgrowth of a course I teach every year at Rutgers, History of the Earth Systems, but I didn't want to write another textbook. I wanted to reach a wider readership and to help elucidate what we know, and perhaps more important, what we don't know, about the role of microbes in making Earth a habitable planet. The lion's share of the work was done while I was on sabbatical at the Radcliffe Institute for Advanced Studies at Harvard University. I am extremely grateful to my hosts at the institute and my fellow Fellows for reading and commenting on the first few chapters. I especially thank Ray Jayawardhana (Ray Jay), Tamar Schapiro, Benny Shilo, and Alessandra Buonanno for helping me in the early days. I am indebted to my friend and colleague, Andy Knoll, at Harvard, for encouraging and critiquing my early chapters and for many discussions while I was in Cambridge, Massachusetts. I thank my late friend, Toshiro Saino, who invited me to give a series of lectures at Nagoya University in 2006. Early on, the lectures in Japan helped me focus on structuring this book. Conversations with many people over the years helped me shape my ideas about the role of microbes in the origins of life on this planet. I thank Tom Fenchel and Ed Delong for collaborating with me on a paper describing the role of microbes in maintaining biogeochemical cycles; the basis of that paper was critical in helping me develop several chapters in this book. The late Lynn Margulis was extremely supportive, and we had many discussions over dinner about symbioses. Joe Kirschvink and Minik Rosing helped me understand what tales very old rocks can tell. Many people were extremely helpful in reading chapters and giving constructive comments; I am especially grateful to Sam Elworthy while he was at Princeton University Press for soliciting the book and to my editor, Alison Kalett, for her patience and insights that helped to improve it. I especially thank my wife, Sari Ruskin, for her very constructive comments and her loving encouragement. My longtime friend Bob Kross gave me many thoughtful suggestions. I thank Ford Doolittle, Dave Johnston, Don Canfield, Paul Hoffman, and Doug Erwin for pointing out key errors that otherwise would have gone unnoticed. Nick Lane was extremely charming in his praise of the few chapters I asked him to comment on, and I delighted in talking with him about the basic concept of the book. Many of my students, postdocs, and collaborators down the years have helped shape my thoughts about the role of microbes in the evolution of life. I am grateful for the support of my research from NASA, the National Science Foundation, the Agouron Institute, and the Gordon and Betty Moore Foundation. I thank and apologize to Sari and our two daughters, Sasha and Mirit, for their understanding and patience and the time stolen from them while I was writing. I thank my colleagues at Rutgers University, where I have worked since 1998. I never would have guessed that as a biophysicist and oceanographer I would one day wind up teaching History of the Earth Systems in a geology department. But, most of all, I thank my parents, neither of whom was a scientist, for encouraging me as a child to follow my dreams in life and for providing me with the intellectual opportunities and emotional support that have served me so well throughout my adult life.

LIFE'S ENGINES

|||||||||

Prologue

Life is a series of connected historical accidents, contingencies, and opportunities. I grew up in a New York City housing project at the edge of Harlem. When I was about nine years old, my mother befriended a young couple in the building. They were graduate students at Columbia University and lived a few floors below us.

Bill Cohen and his wife, Miriam, were studying biology and kept tanks of tropical fish in their apartment. They seemed like a wonderful young couple, and my mother surely advised them on things they didn't need to know. Regardless, they didn't have children yet, and shortly after my mother introduced me to them, they invited me to visit their apartment and see their aquaria. I was hooked.

A few weeks after our introduction, Bill and Miriam gave me a small aquarium, and I started to grow guppies and a green alga, *Nitella*, and watched as the gravid females gave birth to new guppies in the algal bed. I began to read everything I could about tropical fish and became increasingly obsessive about them, and fish in general. I was, unwittingly, on my way to becoming a biologist—all because of a chance encounter by my nosey, loquacious mother with a couple of graduate students in an elevator.

As time went by, I saved most of my allowance and the money I got from doing small jobs and bought more and larger aquaria and increasingly expensive, exotic fish from the legendary Aquarium Stock Company, which spanned a whole city block between Warren and Murray Streets in lower Manhattan. This was the place where addicts of tropical fish satisfied their habit.

About the same time, my father bought me a small microscope at the American Museum of Natural History, which we visited together virtually every Saturday for several years. The microscope was a lot of money for my father, and it almost certainly cost too much, but I had been yearning for it for a long time; it was a birthday present that changed my life. I know museums have to charge for things like microscopes, but it would be much better if they could just give them away to every child who visits.

My father's gift allowed me to see and explore the invisible, magical world of microscopic organisms swimming in my fish tanks. Even though the microscope wasn't high quality, it gave me access to a world I could not have imagined. The organisms were beyond amazing.

I spent hundreds and hundreds of hours looking down the barrel of the microscope, trying to understand the surreal microscopic world that was playing out before my eyes but was so foreign to my personal experiences. I could observe microscopic organisms ingesting smaller particles. I could see single-celled organisms dividing. I could see organisms swimming and others moving by "walking" on the slides. I didn't understand how these organisms moved, how they ate, or how they lived.

By reading books borrowed from the local public library on 125th Street, I started to learn about the microbial world. The library also had an inspiring wooden model sailboat on the imposing staircase that led to the first floor. To get to the adult section, where the science books were kept, I had to pass the sailboat. Between the sailboat and the science books, I could dream about worlds beyond Harlem. I became increasingly absorbed with learning about the exotic places in Africa and South America where my fish came from and what microbes I could identify from drawings in the few books the library had on that subject.

With my microscope and books from the library, I started to understand how paramecia used their cilia to move and how amoeba glided over surfaces in the fine-grained gravel that served as the sediment in my tanks. I got to see that some organisms are attracted to light and others are not; that some organisms required light for their livelihoods, but others required the addition of organic matter. I started to grow microbes from samples of water I collected from the lakes in Central Park and puddles on Riverside Drive. I tried to "think" like a microbe, which as a child, is not so hard to do, even if it is in your imagination.

As the fish bred in my tanks, I could study the development of their embryos in their transparent egg cases. With my microscope, I could see the shapes of the various algae growing on the walls of my fish tanks and how the snails scraped the algae off and ingested them. When I stirred up the gravel or rearranged the rocks in my tanks, I could see, on slides, all the detritus and barely make out the movement of the smallest microbes, which everyone called bacteria. At that time I didn't really understand what these "bacteria" were or their relationship to the fish and plants in my aquaria.

My mother, who was perpetually paranoid about food poisoning, always warned me about "germs" in my aquaria that would make me sick if I drank that water. I didn't really understand what germs were, but I knew they were bad. She had me wash my hands after I rearranged rocks or took samples. I surely wouldn't drink the water that my fish were living in, but it puzzled me why I would potentially get sick if I did.

The fish in my aquaria didn't get sick from the germs and surely they were drinking the water, or so it appeared. Would I really get sick from drinking the water in my fish tanks? I didn't dare—but the water originally came from the faucet in the bathroom of my apartment. I drank water from the faucet every day. But if I used that water straight out of the tap to grow fish—they died. I knew the fish couldn't tolerate the chlorine in the water that came directly from the tap and that they didn't thrive unless they were in an environment where there were bacteria and other microscopic organisms. Yet I could drink the water with the chlorine but would almost certainly get sick if I drank the water from my aquaria. How could I live in a world where chlorine in water was safe to drink, whereas my fish could die if I exposed them to the chlorine that killed the germs in their world? That didn't make sense.

Microscopic organisms appeared to be both good and bad. It was not easy for me, a nine-year-old, to understand that apparent paradox. The germs that so terrified my mother appeared to be important in my aquaria. I became increasingly aware that germs were microbes. At the time, no one knew that all of us have many, many microbes in our guts and that they are as important to our lives as the microbes in my fish tanks were to my fish.

I became more and more fascinated, if not obsessed, with the world of microbes. I spent countless hours, late at night, looking down the barrel of my microscope at samples from my aquaria, and listening to Cousin Brucie playing the 1960s hits on WABC in the earphone of my crystal radio.

For several years, my life was totally absorbed with my fish tanks, my microscope, and the microbes in my fish tanks. But when I was about thirteen, I began to branch out. I became increasingly interested in another, invisible, world—electromagnetic radiation. I didn't call it that then. I think I just called it radio waves—or something to that effect. How did images and sound get transmitted from a station far away to my apartment? That phenomenon seemed beyond incredible.

My parents were electronic Luddites. They were no help in understanding the radio, let alone television. We listened to the radio as a family—but only classical music (my parents were not into jazz or rock and roll). We didn't have a television. My father called television a time thief and thought it was totally irrelevant to life. We literally had thousands of books in our house—and my father read and read and read. He made sure that I knew how to read serious literature. Were he still alive, I am not sure what he would call the Internet, probably something like a time extortion mob. Yet, somehow, although he instilled in me a great respect for literature and the written word, as I watched television in friends' houses, I became interested in learning how sounds and pictures could be transmitted across space without wires. For me, the sounds and pictures were transformative. I couldn't imagine how they could be transferred across the ether to be played on a television, but I might possibly understand exactly how Cousin Brucie got to play a record somewhere in midtown Manhattan and I could hear it several miles away on my crystal radio. I set out to learn how that magic worked.

I had bought cheap electronics parts in small stores down on Canal Street and made a crystal radio. The strongest signal was 770 AM WABC. In fact, it was so strong, it was the only one I could listen to on my crystal radio, which used the incredibly small electrical field generated by radio waves as its power supply. I could attach an alligator clip from my crystal radio to my radiator and listen for free through a small earphone.

Cousin Brucie was a hyper disc jockey who shouted a blurb for the next song and told you who was hot. Totally cool—Brucie became the guy to listen to while cleaning and arranging the rocks in the aquaria.

As I grew, I worked odd jobs in the neighborhood and made enough money to buy very exotic fish for my aquaria. I also bought used and surplus electronic components in the myriad shops on Canal Street. I became an aficionado of African cichlids, while simultaneously building amplifiers, radios, and other simple electronic equipment. I learned basic genetics though breeding and selling exotic fish to Alfred at the Aquarium Stock Company. I learned how electrons could be slowed down by resistors, trapped in capacitors, how electronic tubes worked, and, by building radios and small transmitters, how invisible radio waves were sent and received. But in the back in my mind, I remembered the model sailboat in the library on 125th Street. It was a beacon to a world beyond.

It took another twenty years before I really appreciated how the organisms we cannot directly see with our eyes transformed our planet by developing a global electronic circuit of life. They silently move electrons, but their electronic circuit is not a metaphor; it is truly the engine of life on Earth. Although they were not on display in the Museum of Natural History, they created the gases that allowed me to live. They removed my waste products. They made this speck of dust in the galaxy a habitable planet.

Later in my life, the world in the aquaria that I could see with the microscope my father bought became increasingly important to me, but I didn't know exactly why. It took me several decades to understand that the death of the microscopic organisms and their decay in the gravel in the aquaria of my childhood were miniature models of how organic matter could become the fuel for the car I drive. Over the course of my scientific life, I began to understand that the electronic circuits I built as a child were analogues of life, but they were incomplete. Something was missing. I realized I didn't understand key mechanisms about how cells function. They don't obtain energy from radio waves; they get energy from higher-energy particles of light emitted from the Sun. More puzzling, unlike radios, which don't grow and develop from radio eggs to make new radios, cells assemble and replicate themselves, time after time. The replication of cells is one of life's most critical functions.

The tension between replication and metabolism remains one of the most difficult hurdles in understanding how life evolved on Earth. It requires better knowledge of the electronic-circuit diagram of life. The two worlds were not readily connected in my mind. To be honest, I also didn't pay much attention to invisible worlds in my formal education. Connecting the world of an electronic circuit of life with the evolution of organisms wasn't exactly the vision or mission of my teachers in high school or my professors in college. I had to discover the connections for myself.

I attended a high school where biology was an optional course in an area I wasn't studying. I was drilled in math, physics, and chemistry. It wasn't until much later in life that I realized that books on biology, which were assigned to me in college, mostly ignored microbes, except as carriers of disease ("germs"). Discussions on evolution, when there were such, almost always focused on animals and plants. The biology texts I was required to read were not only inaccessible, they were also downright boring. I couldn't understand how one could take such an exciting subject, the study of life, and turn it into something so filled with irrelevant jargon.

Regardless, as a college student in New York thinking about the world in which I lived, I remembered seeing many butterflies in the park nearest my home—along Riverside Drive. From an article I read in *National Geographic*, I distinctly recalled a discussion of the migration of butterflies from an obscure place in Mexico across thousands of miles to the north, to Riverside Park. I could only wonder what they experienced in their migration to this seemingly lost land of Harlem. It was beyond incredible that these apparently delicate animals could sustain a migration of thousands of miles. They were, to me, living emblems of a force of life. Like the dream enshrined in my young mind by the model sailing boat in the library on 125th Street, the butterflies escaped from their boundaries to discover new worlds.

In college, we were shown how to distinguish between the right and left eyes of a cow, the names of the bones in a human hand, and the names and shapes of various flowers and fruits. The evolution of teeth and the developmental stages of chicken embryos were given great weight.

The result was that the ensuing, increasingly unmemorable, and mostly irrelevant vocabulary of biology became more important than the subject itself. In the end, my formal education in college had the not-unexpected result of expunging from me virtually all of the wonders of biology that had inspired me as a child. Wonder gave way to a formalized language and ritualized culture of science. It is a philosophical cult that is ingrained in most aspiring scientists so rigidly that core questions, such as What is life? When did life originate? and How does it work? become distant memories, if ever they were asked in the first place.

Not unlike drill sergeants, many of my professors worked hard to wring these and other irreverent questions out of me; the wonder, let alone the joy, of biology, or of science for that matter, meant nothing to the future of the premed students they catered to. If I was going to be a suc-cessful future soldier in the force of biological research, I had to know the vocabulary and the facts and to forget about the electronic circuits of life and microbes. I do not blame my professors, many of whom had the best intentions. It was, and often remains, a culture in science—find the "best" and weed out the "worst." It is a problem of how to inspire young minds to tackle the most difficult problems—and understanding the origins of life is difficult. Unfortunately, in weeding out the worst, some teachers often seem to systematically eliminate the most inquisitive and creative minds in science.

It wasn't until much later, when I began to work seriously in nature's real aquarium, the ocean, that I started to think about why there are no butterflies on Venus; or if there ever had been, would we ever know? I began to realize the extent to which microbial processes control and make Earth habitable for plants and animals, including us, and how the organisms I once viewed down a microscope as a child are connected to each other by an invisible, yet real, electronic circuit of life. That circuit makes this planet function.

This book is an attempt to explore and explain how the global elec-tronic circuit came to exist, how it controls the balance of nature on Earth, and how humans can disrupt it, to their potential peril. Let's begin with what we see, and often don't, in the macroscopic world in which we live.

CHAPTER 1

|||||||

The Missing Microbes

A few years ago, I was given the opportunity to work on a research ship on the Black Sea off the north coast of Turkey. The Black Sea is a fascinating and unique body of water: below the upper 150 meters or so, there is no oxygen. The focus of my work was to study the photosynthetic microbes in the upper 150 meters.

Photosynthetic microbes use the energy of light from the Sun to make new cells. Throughout the world's oceans, there are microscopic photosynthetic organisms, the *phytoplankton*, that produce oxygen. They are the forerunners of higher plants but evolved much earlier in Earth's history. After several days, the instrument my research group used to detect phytoplankton, a special type of fluorometer that we had developed years earlier, recorded some strange signals that none of us had ever seen. The signal was quite deep in the water column: just at the location where all oxygen is gone and the light intensity is very low. As we worked, I realized that the organisms responsible for the strange fluorescence signal occupied a very thin layer, perhaps only a meter or so thick. They were photosynthetic microbes, but unlike the phytoplankton higher up in the water column, they could not produce oxygen. These microbes were representatives of an ancient group of organisms that evolved long before phytoplankton. They were living relics of life at the time before there was oxygen on the planet.

Working on the Black Sea had a profound influence on how I think about the evolution of life on Earth. In my mind, sampling deeper into the water column was like going back in time to find microbes that had once dominated the oceans and are now confined to a very small fraction

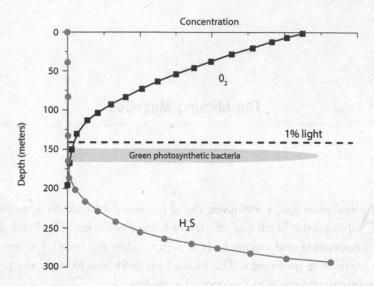

FIGURE 1. An idealized profile of dissolved oxygen and hydrogen sulfide gas (which smells like rotten eggs) in the upper 300 meters of the Black Sea. This body of water is unique in the ocean; in most ocean basins and seas, oxygen is detectable to the seafloor. Just below the depth at which 1% of the sunlight from the surface remains, there is a very narrow layer of photosynthetic bacteria that split the hydrogen sulfide with energy from the Sun, for their own growth. The metabolism of these organisms is extremely old; it probably evolved more than three billion years ago, when oxygen concentrations on the Earth's surface were extremely low.

of their former habitat. The photosynthetic green sulfur bacteria, which turned out to be the organisms responsible for the strange fluorescence signal, are obligate anaerobes; they use energy from the Sun to split hydrogen sulfide (H_2S) and use the hydrogen to make organic matter. These organisms can live at very low light intensities but cannot tolerate exposure to even small amounts of oxygen.

As we traversed the Black Sea over the next several weeks to sample different areas, we saw dolphins and fish in the upper ocean, but there were no multicellular animals below the upper 100 meters or so. Animals can't live for long without oxygen, and there appeared to be none in deeper waters. Microbes had altered the environment of the Black Sea. They produced oxygen in the upper 100 meters but consumed the gas further down. In so doing, they made the interior of the Black Sea their exclusive home.

After about a month at sea, I found myself back in port in Istanbul, admiring Turkish rugs. Mount Ararat, in northeast Turkey, is famous for its woven rugs depicting the story of Noah's ark. The kilims from that region are rich tapestries with pairs of giraffes, lions, monkeys, elephants, zebras, and all sorts of familiar animals woven into them. As the merchants unrolled their wares and provided endless cups of sweet tea, I started to think about how the story of the ark has influenced our distorted understanding of life on Earth. On one hand, the story is about destruction and resurrection. On the other hand, it is about how God made humans stewards of life. In neither case do microbes appear as creators nor destroyers of life.

The word "evolution" literally means "to unroll," but as the merchant unrolled the beautiful rugs for me, I began to see how the biblical story of the ark failed to provide a clue about how life evolved. Was all life on Earth preserved by Noah and taken on the ark? Could some organisms have been left behind? Although the story of the ark is deeply embedded in Western culture, it fails to inform us about the origins of life. To begin to understand the origin of life requires another perspective, one based on science and, especially, its application to the evolution of microbes.

To a large extent, science is the art of finding patterns in nature. Finding patterns requires careful observations, and inevitably we are biased by our senses. We are visual animals, and our perceptions of the world are based primarily on what we see. What we see is determined by the tools we have. The history of science is closely tied to the invention of novel tools that allow us to see things from different perspectives, but ironically, the invention of tools is biased by what we see. If we don't see things, we tend to overlook them. Microbes were long overlooked, especially in the story of the history of evolution.

The first few chapters in the ongoing story of the evolution of life on Earth were written largely in the nineteenth century by scientists who studied the fossils of animals and plants—fossils they could easily see. The patterns in nature they observed ignored microbial life for two simple reasons: there was no obvious fossil record of microbes in rocks, and the pattern of microbial evolution could not be easily discerned by looking at living organisms. The tools for finding fossils of microbes barely existed; however, even if there had been such tools, the roles these organisms played in shaping the evolution of Earth would not have been appreciated until new tools became available in subsequent decades. The patterns of evolution observed in animals and plants were historically inferred from the shapes and sizes of their fossils and the arrangement of these fossils through geological time. That approach does not work nearly as well when applied to microbes.

In sum, the oversight of microbes, in both the literal and figurative senses, distorted our worldview of evolution for more than a century, and including microbes in our understanding of evolution is still a work in progress. In as much as science is the art of discovering patterns in nature (and that is difficult enough), it is also about discovering patterns that we cannot see with our naked eyes.

But first, let us briefly examine the story of evolution as it emerged in the nineteenth century. That was when many of our scientific concepts about life came to be formed. The ideas were largely based on what could be seen and framed by biblical stories of the creation, including the flood and the story of Noah's stewardship of God's animals, such as the stories woven into Turkish rugs.

In the early 1830s, a gentleman scientist, Roderick Impey Murchison, and a charismatic Cambridge University professor, Adam Sedgwick, reported that there were fossils of animals deep in the ground at a site in Wales. Fossils had been known for centuries, but their significance was not well understood. Many people realized that these were impressions of organisms that had died long ago—but how long ago was not clear, nor was it clear how the impressions were preserved.

Sedgwick was one of the foremost authorities on fossils in Britain, and one of the students who attended his lectures was Charles Darwin. In the summer of 1831, barely yet twenty-two years old, Darwin went with Sedgwick into the field in north Wales to learn about fossils firsthand.

That experience changed Darwin's life forever. Darwin not only helped Sedgwick find fossils of animals in the rocks, he also learned basic principles of geology, and those observational skills would be very useful to him for the rest of his life.

The fossils found in the rocks in England and Wales by Sedgwick and Murchison were also found elsewhere in Europe, and a system of classification based on the sequences of the fossils in the rocks began to take hold. Often the physical appearance of the fossils resembled familiar animals that lived in the oceans, such as clams, lobsters, and fish; however, some of the fossils were incredibly bizarre, and no one had ever seen anything like them in the oceans of the times. There was tremendous controversy about the meaning of the fossils, but the discoveries clearly suggested a sequence of changes in animal forms from lower to higher levels in the layers that these ancient marine sediments formed. At the time, it was generally understood that rocks deeper down in a sequence were older than the rocks above.

The discovery of animal fossils in rocks was hardly new. Probably the most famous early description of fossils was recorded by a Danish scientist, Nicolas Steno, in 1669. He had found objects that looked very much like shark's teeth in rocks in Italy, but how objects from once-living organisms could be so preserved puzzled him. Steno, however, thought carefully about how the fossils were arranged in the rocks. They were arranged in layers, and it appeared to him that the older layers lay below younger layers. This notion, called *superposition*, is one of the primary rules in sedimentary geology and strongly influenced the interpretation of the fossil record by Sedgewick more than a century later. Steno eventually abandoned science and entered the Church to devote his life to God. His early work on fossils was largely forgotten, and he himself believed that life began as it was described in Genesis.

To me, the logic of the idea that the fossils preserved in rocks are arranged in some accordance with time was an extraordinary insight, but it was not easily supported, because basic geological information was not yet available. To a large extent, the effort of finding patterns in fossils awaited the great mind of Charles Lyell, one of Darwin's intellectual mentors and a close friend. Lyell, a Scottish barrister turned naturalist, is often credited with founding a new area of science, which he called *geology*. Lyell, like Steno, realized that there was a logical sequence in the fossil record; however, unlike Steno, Lyell expounded on geological processes, such as erosion, volcanism, and earthquakes, to help explain the sequences observed in the fossil record. Indeed, his elucidation of the fossils in the rock sequences would later inspire Darwin to muse upon how organisms change over time. The lifelong friendship between Lyell and Darwin was a legendary symbiosis in science.

On December 27, 1831, as Darwin was beginning his voyage on the HMS *Beagle*, a ninety-foot, ten-gun brig with seventy-four people on board, he was allowed to have very few books in the very cramped chart room, which was his assigned sleeping quarters. He slept in a hammock in the 9- by 11-foot room, which had a 5-foot ceiling; it was dark and uninviting, and he had to share the quarters. Among other things, he took with him the first volume of the first edition of Lyell's new book, *Principles of Geology*, which had been published in 1830. He also took his personal copy of the King James Bible. On ships I work on, I have a hot shower every day, and while I sometimes share a small cabin, there is a library on most research vessels. Perhaps, then, it should not be too surprising that Darwin used seasickness as an excuse to leave the *Beagle* at almost every opportunity and to wander across the continents to meet the ship at another port of call.

Lyell took on the hard task of explaining to an interested public how animal fossils could wind up in the Alps in central Europe, as well as in the hills of Scotland and throughout the British Isles. One of the basic problems was time and how the Earth came to be formed.

Several arguments had been put forth over the centuries. One, from medieval times, was that that God made rocks to look like familiar organisms to test the faith of his flock. As absurd as it is, the notion still has many proponents, especially in parts of the United States. A second idea was that in ancient times, volcanoes exploded and carried animals from the oceans onto land, where they died and their skeletons were preserved in the rocks. A third concept was that the animals died after the Great Flood, when the sea level dropped. Indeed, this diluvian origin of fossils appealed to Sedgwick himself. There were several other ideas, which Lyell recounted eloquently and with precision, as a barrister might present a case to a jury.

Lyell proposed the radical idea that the fossils from marine animals were found in rocks on land because a long time ago the rocks were under water. Over time, the rocks were somehow uplifted and deposited on land. That notion, tested many different ways, is actually correct, but the processes responsible would not be uncovered until more than a hundred years later. One of the major problems Lyell faced was accounting for the age of the Earth. How long was "a long time ago"?

The age of the Earth had been meticulously calculated by James Ussher, the Archbishop of Amargh, in the book *Annales Veteris Testamenti*, which was published in 1654. It was taken by virtually every educated British citizen as the most accurate estimate of the time of creation. On the basis of a literal interpretation of the Bible, Ussher had determined that the Earth was formed at nightfall of the Sunday preceding October 23, 4004 BCE in the Julian Calendar; that is, about 6000 years ago.

As a student of law, Lyell had been trained in argumentation and was amused by some of the illogical and sometimes irrational thought processes used to explain the existence of and changes in fossil animals. He understood the power of argumentation and wrote that "the system of scholastic disputations encouraged in the Universities of the middle ages had unfortunately trained men to habits of indefinite argumentation, and they often preferred absurd and extravagant propositions, because greater skill was required to maintain them; the end and object of such intellectual combats being victory and not the truth." But even talented barristers can't win arguments against the written word of God.

Lyell didn't know anything about how evolution might work, let alone how to measure geological time. He thought Jean-Baptiste Lamarck's theory—that traits were acquired by animals during their lifetime and somehow passed on to future generations—was as good as any and more rational than most. Indeed, Lamarck's work on animal forms (he was the world's leading authority on animals without backbones—the invertebrates) led to him to propose that organisms could be arranged along a chain in time, from the simplest to the most complex forms. Lamarck set in motion the idea that organisms somehow change—that is, evolve—over time. Indeed, although now largely unjustifiably ridiculed or ignored in biology texts and classes, Lamarck was the intellectual father of a field he called *biology*.

The idea that fossils of animals were arranged in layers of rocks along an arrow of time got Darwin thinking about life on time scales he could barely imagine and could not easily quantify. If the oldest fossils were many meters beneath other fossils, how long had it taken for the layers of rock to build up?

Darwin was extremely puzzled by the early fossils that Murchison and Sedgwick had found. He knew that beneath the layers of rocks with fossil animals were layers that contained no fossils, but he could not understand why. The record of animals appeared to come out of nowhere, and their evolution appeared to be relatively rapid. But how rapid? And why, all of a sudden, were there fossils of fish, but in the rocks below there were only organisms that looked like invertebrates? And even further below, why were there no fossils of animals at all? It was the geological equivalent of unrolling a Turkish rug depicting the story of the ark, but half or more of the rug had no animals. Darwin needed to explain these issues first to himself and then to his colleagues. To answer these questions, he needed to try to date the rocks, and for that he needed a clock.

On September 7, 1859, the bells in the clock tower housing Big Ben rang for the first time. The clock was meticulously designed and is extraordinarily accurate; indeed, it is an iconic symbol of English engineering and craftsmanship at the dawn of the Industrial Revolution. Two months after that historic event, on November 24 to be precise, John Murray, III, the venerable London publisher on Albemarle Street, sent Charles Darwin's new book, *On the Origin of Species by Means of Natural Selection, or the Preservation of Favoured Races in the Struggle for Life*, to market.

In Chapter 9 of *The Origin of Species* (the title was later shortened), Darwin attempted to account for the time required for extinct fossil animals to have changed, or evolved, to become the modern forms. The problem was not straightforward. Lyell and his predecessor, the Scottish physician James Hutton, had proposed that the Earth was infinitely old. Darwin did not know whether that concept was true, but he certainly believed that it had to be more than 6000 years old. To obtain a more realistic age, he developed a rather interesting, if not downright ingenious, approach to measuring geological time.

Darwin's clock was based on a geological phenomenon: the rate of erosion of sedimentary rocks, the kind that contain fossils. He specifically chose the Weald, a well-studied chalk and sandstone cliff abutting the sea in Kent, England. Darwin calculated that this formation eroded about one inch per century, and based on the size of the formation at the time, he calculated that the "denudation of the Weald must have required 306,662,400 years; or say three hundred million years."

Darwin didn't account for the time required for the formation of the cliff itself, but that was a detail. Moreover, he didn't account for the rocks below the Weald, which would have only made the age of the cliff even older, and possibly infinitely old, as thought by Lyell. Darwin's estimate of the age of the cliff certainly was a bold speculation, and without a better constraint, it was apparently based on a rational, physically verifiable idea. The implication was obvious. The Earth was very old. It was much, much older than Ussher had calculated, and it was a lot older than most people could possibly have imagined at the time. And while the date of the origin of life on Earth had not been determined (and remains unclear to this day), that there were rocks that had no fossils beneath those above implied that Darwin's estimate of the age of the Earth was conservative.

Regardless, millions of years are not the history in the Bible, and they certainly didn't fit what everyone had been taught at school. Darwin clearly knew his estimate was going to be met with skepticism, but he had no way of knowing what was to come. Besides assaulting the Biblically held, seventeenth-century calculations of the Archbishop of Armagh, Darwin's estimated age of the Earth was assailed by a fellow scientist, the Einstein of the day, the physicist William Thomson, later to become Lord Kelvin. Thomson set out to put the record straight, based on first principles of physics.

Thompson argued that the age of the Earth could be accurately determined by assuming that the planet began as a molten rock and subsequently cooled. Using measurements of the change in temperature with depth through the Earth's crust and experiments he performed about the conduction of heat by rocks, he developed an equation for how fast the Earth had cooled to its present state. In 1862, Thomson proclaimed that the Earth was about 100 million years old. He admitted a huge uncertainty of between 20 and 400 million years, but as time went on, he became increasingly dogmatic and convinced that the actual age was closer to 20 million years. This estimated age appeared to be too short to allow evolution, as Darwin envisaged it, to proceed. Thomson became one of the harshest critics of Darwin's new ideas regarding evolution, not because he did not believe in evolution per se, but rather, because as a physicist, he did not believe the calculations of the age of the Earth based on geological processes such as rates of erosion. Ultimately, the contrarian views of Thomson forced geologists to develop better models for the age of the Earth, but doing so would take almost another century.

If Darwin was even remotely correct, then life evolved on Earth over a very, very long time—much, much longer than anyone imagined. But how did it evolve? In a doodle on page 36 in Notebook B from 1837, Darwin sketched a tree of life in which he had the radical idea that organisms were related to each other from a common ancestor and that their relationship could be discerned from similarities in physical appearance. That basic notion was similar to Lamarck's concepts that had been developed more than fifty years earlier; however, Darwin had a different idea as to how the process occurred.

The changes in the animal forms were subtle and, based on the distance between fossils in the rock record, appeared to be slow. In addition, for the proposal to work, some organisms that appeared earlier in the fossil record had to go extinct and be replaced by new species, or the Earth would have an ever-increasing number of animal and plant species. In other words, once an organism becomes extinct, it should never reappear later in the fossil record.

Darwin realized that this remarkable, revolutionary idea would be challenged; and so it was. The fossils were clearly relics of animals and plants, but there were no bones of humans in the rocks. If that were true, then Darwin clearly understood the implications of the "missing" humans; like animals in the fossil record, we must also have arisen by some process that allows one organism to evolve into another over some undefined, but prolonged, time.

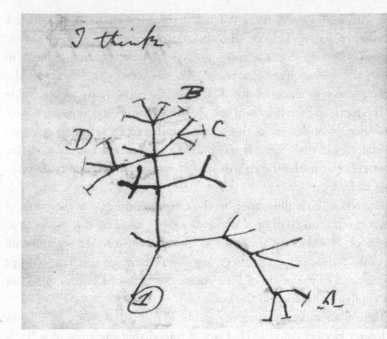

FIGURE 2. A reproduction of the doodle that Darwin sketched in Notebook B between 1837 and 1838. The basic idea is that extant species are descended from extinct species but are also related to other extant species to form a historical tree of life. This doodle was the kernel for the theory of descent with modification followed by selection—the core of Darwinian evolution. (With permission from Cambridge University Press and thanks to Peter and Rosemary Grant. Copyright © 2008 The Committee for the Publication of Charles Darwin's Notebooks.)

The concept of genes and the basis of the physical inheritance of traits were totally unknown to Darwin, or anyone else at the time. (Gregor Mendel would not present his work on inheritance of traits in peas until more than six years after the publication of the first edition of *The Origin of Species*, in 1866). Indeed, despite the confusion in most biology texts, Darwin would not have had a major problem accepting Lamarck's basic concept that organisms can inherit traits from their environments; however, Darwin's major contribution was the idea that within all species there is natural variation that can be selected. Breeders of dogs and pigeons did this all the time. However, in nature, Darwin proposed that the traits are selected by the environment in which the species lives. Selection either enhances the ability of the organism to reproduce, or not. If it does, then the traits most suitable for the particular environment are passed on to subsequent generations. The concept of descent with variation followed by selection occupies six chapters in the *Origin*. It was one of the most remarkable scientific ideas ever put forth, and to this day, it remains a core, unifying principle of biology.

There is a single illustration in the *Origin*, at the end of the book, of a hypothetical origin of taxa, which is loosely based on the doodle from Notebook B. Curiously, the figure does not show a single origin for all taxa but, rather, many origins giving rise to new species. The concept of origin, as in the origin of life, was in the back of Darwin's mind but not explicitly discussed in the book.

More than a decade after the publication of the *Origin*, in a letter to Joseph Hooker dated 1871, Darwin mused that life arose in a "little warm pond with some sorts of ammonia and phosphoric salts,—light, heat, electricity, etc. present, that a protein compound was chemically formed, ready to undergo still more complex changes, at the present day such matter would be instantly devoured, or absorbed, which would not have been the case before living creatures were formed."

Eighty years after that notion was put forth, a young chemist, Stanley Miller, and his Nobel Laureate advisor, Harold Urey, actually did make amino acids (the building blocks of proteins) in a laboratory at the University of Chicago. They used ammonia gas, methane, hydrogen, and water and a spark to simulate lightning. That experiment, which was published in 1953, gave great hope that an understanding of the origin of life was imminent. However, there is a huge gap between making the chemical constituents of organisms and making the organisms themselves. In even the simplest organisms, the chemical constituents are organized into microscopic machines that give rise to metabolic processes and allow the cell to replicate. No one has yet created a living organism from scratch, but that is not to say it is impossible.

The simplest organisms are microbes, organisms of which Darwin was surely aware but not certain how to accommodate in his theory. Indeed, Darwin took a microscope with him on the *Beagle*. (Along with his Bible and natural history books, he also took two pistols, twelve shirts, two books to help him learn Spanish, and a coin purse.) But because microbes do not leave a fossil record that is clearly visible to the naked eye, Darwin could not have known that the rocks beneath the visible fossils were not from a period in Earth's history that was before the origin of life—but rather a period before animals or plants. Even if Darwin had observed fossil microbes, he would almost certainly not have understood what their relationship to plants or animals was. Darwin, and virtually every other scientist in the nineteenth century, would have been profoundly surprised to learn that plants and animals had all descended from microbes over a period of time that was completely unimaginable in the nineteenth century—far longer than 300 million years. Indeed, microbes are not mentioned in the Bible, except indirectly in reference to diseases like the plague. They certainly weren't deliberately taken by Noah on the ark, nor are they woven into the Turkish tapestries depicting the story of the Great Flood.

While we have made great progress in the 150 years since the publication of the *Origin*, scientists are still struggling to understand whether life began in a small warm pond, at a deep-sea hydrothermal vent, or somewhere else. How might it have started? How did it get going? How did microbes lead to the evolution of plants and animals? How were these organisms missed for so long in our search for the origins and evolution of life?

The answers to these questions are complex, and many aspects are still far from fully understood, but we have learned a lot because of the tools developed during the last century. If Darwin had been on an oceanographic research voyage in the Black Sea early in the nineteenth century, he might have observed that there were no animals below the upper hundred meters and concluded that the deep water was lifeless. But had he been a microbiologist, our understanding of the origins of species would have been very different. Although microbes were well known in the nineteenth century, it took another century before they were included in our understanding of the evolution of life on Earth. Microbes were missed because of our observational biases. They had been on this planet for billions of years before the first animal arose.

Let's meet the missing microbes and see how they played an outsized role in making this planet function. Without microbes, we would not be here.

CHAPTER 2

||||||||

Meet the Microbes

Perhaps one of the biggest ironies in biology is that microbes, which are the oldest self-replicating organisms on Earth, were among the last to be discovered and have largely been ignored. The history of their discovery is, like many in science, based on the invention of new technologies; in this case, the microscope and gene sequencers. The lack of attention to these organisms is largely the result of our own observational bias—we tend to ignore what we cannot see. That predisposition allowed us to make great progress in astronomy, observing visible objects hundreds of billions of miles away from us, long before we could appreciate the role of microbes on this planet. Let's briefly examine the history of the discovery of microbes in the context of our biases.

In the fourteenth century, crude lenses (which were named after the lentil bean because of its double convex shape) were being fabricated in Europe for correcting vision. Simultaneously, artists had begun to develop methods of projecting images on a canvas with simple camera obscura techniques. A camera obscura does not require a lens. It is a box, or even a small room, with a hole that allows light to enter, and an inverted image of the external world is projected on the back of the box. Inside the box, one can follow the rays of light. By tracing the rays and experimenting with glass lenses inside the box, early instrument makers began to understand how to design lenses.

Toward the end of the sixteenth century, the Dutch began to work with Italian glass manufactured in Venice. At the time, Venetian glass was the most expensive because it was the clearest and the highest quality available; with it, the Dutch began to fashion relatively high-quality lenses. Early in the seventeenth century, two Dutch lens makers fashioned a telescope by pairing a concave and a convex lens in a tube. Although the instrument was not much more than a crude spyglass, having a magnification of about seven- or eightfold, it was a huge breakthrough in technology at the time. To this day, lens designers use the same basic formulas that were developed from the ray tracings of these pioneers in that new field, optics.

In 1609, Galileo Galilei, using a telescope made in Italy from a Dutch lens maker's design, observed that the moons of Jupiter orbited that planet rather than the Earth. Although Galileo's instrument had a magnification of only about twentyfold, it was sufficient to allow him to zoom in on what we already could see with our naked eyes: planets, stars, and the moon. His observations threatened the prevailing Ptolemaic, or geocentric, understanding of the importance of Earth in relationship to the rest of the universe, which was that the Sun and all the planets orbited the Earth, and not vice versa. But Galileo exposed us to something more fundamental than stargazing. He showed us a place we did not know, one that made us lesser. Earth became only one planet among several in our solar system. Galileo clearly knew how profound his discovery of moons that orbit Jupiter was. He changed the way we think about our planet, ourselves, and our special relationship to the universe (and hence, our special place in the eyes of God).

Although stories of Galileo and the telescope abound, a somewhat lesser-known fact is that he also had developed a microscope. It had been known for several years that by simply inverting a telescope with two lenses, one could magnify objects nearby. You can do this at home by simply looking the wrong way through the barrel of a pair of binoculars and holding an object, such as the tip of your finger, up very close to the lens. (It's a great dual use of binoculars on a field trip.)

Galileo's microscope, which was developed around 1619, was simply an inadvertent outgrowth of the invention of the telescope. The optical design of the telescope was inverted and put into a new housing. The microscope was smaller than its counterpart, and the two lenses were set in a barrel made of leather and wood. Regardless, Galileo did not seem to have much interest in what he saw with his inverted telescope. He appears to have made little attempt to understand, let alone interpret, the smallest objects he could observe. In fact, it was so irrelevant to him that it was not until 1625 that it was given the name *microscopio*. Perhaps ironically, during an outbreak of the plague, a microbial disease that is transmitted by flea bites, Galileo drew pictures of fleas he saw under his microscope, but the drawings were not widely distributed, and the instrument languished in Italy and was barely used.

The difference between the telescope and microscope is not simply the configuration of lenses, it is also in human perception and the anticipation of what is seen. While the lack of perception may be partially due to hubris, I think most often it is the lack of looking for patterns in nature in places that are not normally accessible to our limited senses. We can see objects far away with our naked eyes. Comets, meteorites, planets, moons, stars, and even exploding stars can be seen without a telescope, and hence when they are brought closer for inspection with such an instrument as the telescope, these distant objects are not so mysterious, just somewhat so. However, our eyes cannot see something much less than the width of a hair (about a tenth of a millimeter) without the aid of a magnifying device. On the scale of microscopic structures, we are virtually blind. We see the moon with our naked eyes, but not our own cells. We see stars, but not molecules. We see distant galaxies, but not atoms. If we can't even understand that there is a microbial world, why would we look for it?

The discovery of the microbial realm, like so many findings in science, was an accident that changed the world as profoundly as Galileo's observation of the moons of Jupiter. It required a focusing of the mind as much as of an instrument. The breakthrough came in 1665, when the English Royal Society published the first popular science book, *Micrographia* (with the subtitle *or Some Physiological Descriptions of Minute Bodies Made by Magnifying Glasses with Observations and Inquiries Thereupon*).

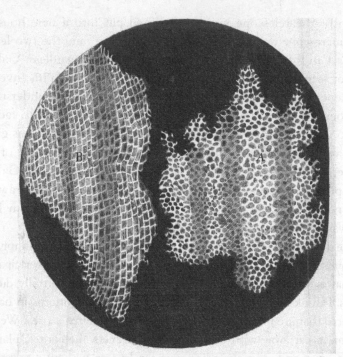

Figure 3. Robert Hooke's drawing of a thin section of cork. He called the structures consisting of pores surrounded by bits of a wood *cells*. This image is reproduced from Hooke's book, *Micrographia*, which was first published September 1665. (© the Royal Society)

The book was written by Robert Hooke, then a thirty-year-old hunch-backed, cantankerous, neurotic hypochondriac who was also a brilliant natural scientist, polymath, and an original Fellow of the society.

Micrographia captured many people's imaginations. In it, along with fifty-seven beautiful engravings based on meticulous illustrations by the author, Hooke provided not only a clear description of his own microscope but also the architecture of fleas (clearly these were as abundant in England as in Italy), the seeds of thyme, the eyes of ants, the internal makings of sponges, microscopic fungi, and the small building blocks of plants. He observed the latter by cutting a small section of a cork with a penknife "sharpened as keen as a razor." In the thin sections of cork he described small structures that looked to him like the rooms in which monks lived. Hooke called these microscopic structures *cells*.

In examining other plants, he understood that cells were ubiquitous and described them in several other species, including fennel, "carrets," burdocks, etc. In the end, *Micrographia* was the first best seller in science. Samuel Pepys bought a copy of the book shortly after it was first published and wrote in his diary, "Before I went to bed, I sat up till 2 a-clock in my chamber, reading of Mr. Hookes Microscopical Observations, the most ingenious book that I ever read in my life." A second edition of *Micrographia* was printed by the Royal Society two years after the first sold out. The book has been reproduced many times since, and it is still in print.

Hooke's observations were based on a relatively simple compound microscope that had two lenses. Instrument makers at the time were familiar with telescopes and designed microscopes with two lenses, very similar to that of Galileo's, because ray tracings clearly showed that these instruments should work well. But two-lensed microscopes had a big, unanticipated problem that telescopes did not. In such simple compound microscopes, the first lens created a halo of many colors, which was then magnified by a second lens. The result was that the more one magnified the object, the more distorted the image became.

The microscope Hooke used was fabricated by Christopher Cock, a very skilled London instrument maker. It was a lovingly crafted, intricately decorated instrument and cost a small fortune, but the optics were poor. It suffered from the large optical aberration that lens makers at the time could not avoid. The best instrument, regardless of how lovingly the fabricator decorated it, could magnify an object only by about twentyfold before it became almost worthless. Even at such low magnification, the images were fuzzy, and sometimes a bit of imagination was required to reconstruct the structure of the object in view. Regardless, Hooke's skillful illustrations were mind-boggling at the time, and publication of *Micrographia* sparked interest in the construction of better lenses.

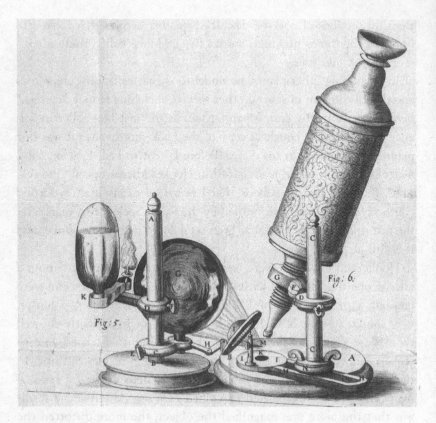

FIGURE 4. An illustration of Robert Hooke's microscope, drawn by Hooke himself and published in *Micrographia*. The microscope, which contained two lenses held in position in an ornately decorated tube, had a magnification of about twentyfold. Light from the Sun or an oil lamp could be focused on the sample by a spherical flask of water. (© the Royal Society)

In 1671, a lifetime after the discoveries of Galileo and thirty-six years after his death, Anton van Leeuwenhoek, a Dutch fabric merchant in Delft, developed a new but far less ornate microscope with smaller, simpler, and, ironically, better optics that allowed much higher magnification without the distortion of the more complicated, expensive instruments. Rather than using two lenses, Van Leeuwenhoek pulled hot glass rods to form threads and then reheated the threads to form small glass spheres.

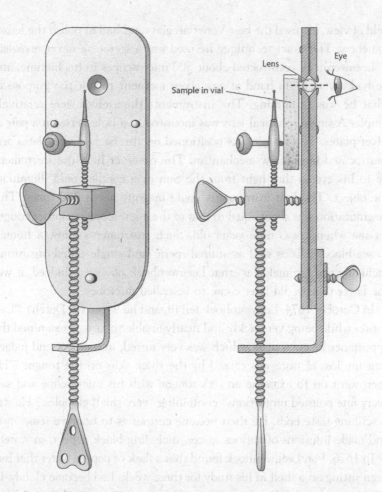

FIGURE 5. An illustration of the type of microscope invented and used by Anton van Leeuwenhoek. The single spherical lens was placed in a small hole between two plates. The sample was held close to the lens with a small screw, and the observer placed his eye close to the lens and held the microscope up the light. Despite its simplicity, this type of microscope could magnify up to 400 times, depending on the quality and size of the lens.

The glass spheres Leeuwenhoek used were about one and a half to three millimeters in diameter. There was a tradeoff in the design of the lens: the smaller the lens, the higher the magnification, but also the smaller the field of view. He used the best Venetian glass and had to polish the lenses somehow. The exact technique he used was a secret he never revealed.

Leeuwenhoek constructed about 500 microscopes in his lifetime, and he had a variety on hand at any given moment to suit the purpose of what he was examining. The instruments themselves were relatively simple. A single spherical lens was mounted in a hole between a pair of silver plates. The sample was positioned on the back of the plates and was focused by a screw mechanism. The observer held the instrument up to his eye so that light from the Sun or a candle could illuminate the object. The best instruments could magnify about 300 times. This magnification was about equal to that of the microscope my father bought for me when I was nine years old. Such instruments allow a human to see blood cells as well as animal sperm and single-celled organisms, including the "animalcules" that Leeuwenhoek observed. Indeed, it was the latter that would later come to be called microbes.

In October 1674, Leeuwenhoek fell ill, and he wrote (in Dutch), "Last winter while being very sickly and nearly unable to taste, I examined the appearance of my tongue, which was very furred, in a mirror and judged that my loss of taste was caused by the thick skin on the tongue." He then went on to examine an ox's tongue with his microscope and saw "very fine pointed projections" containing "very small globules." He was describing taste buds. He then became curious as to how we sense taste and made infusions of various spices, including black pepper, in water.

In 1676, Van Leeuwenhoek found that a flask of pepper water that had been sitting on a shelf in his study for three weeks had become cloudy. In examining the cloudy water with one of his microscopes, Leeuwenhoek was surprised to find very small organisms swimming around. The organisms were only 1 to 2 micrometers in diameter—about one hundredth the diameter of a human hair! He sketched the cells and wrote, "I saw a great multitude of living creatures in one drop of water, amounting to no less than 8 or 10 thousand, and they appear to my eye through the microscope as common as sand does to the naked eye."

The discovery of animalcules was itself unforeseen. It was like seeing moons of Jupiter but without a planet for the moons to orbit. It was a portent of untold numbers of invisible organisms and their presence

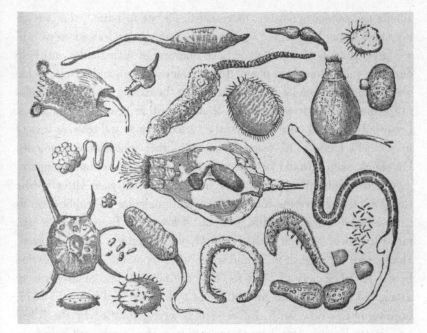

FIGURE 6. Illustrations of animalcules—the microbes discovered by Anton van Leeuwenhoek. In the seventeenth and eighteenth centuries, it was thought that microbes were microscopic animals with heads and stomachs and that their progeny were produced by sexual conjugation between males and females of the same species.

right here on Earth. Leeuwenhoek had no idea what the organisms really were. He imagined they were literally extremely small animals, containing organs such as a stomach and a heart, just like the large animals we see with our naked eyes.

It is truly remarkable that the single-lens instruments made by Leeuwenhoek were capable of allowing him to see organisms so small, yet even with the best lenses of the day he could not resolve their internal structures. However, Leeuwenhoek did something even more profound. Following the discovery of organisms in the pepper water, he examined scrapings from his own mouth. He was astonished to see, for the first time, the presence of animalcules on his teeth and gums. Here Leeuwenhoek really stood out among the natural scientists; he revealed, for the first time, that we are not alone in our bodies. We are carriers of animalcules. Indeed, as we will see later, animals like us harbor huge numbers of animalcules and help distribute them around the planet through our excretions and secretions. He also noted that when he drank hot coffee in the morning, the animalcules in his mouth died; it was the first observation that heat killed microbes. Leeuwenhoek went on to describe the various shapes and relative sizes of microbes he found in his own saliva and in other aqueous environments. His simple sketch would later become the basis for microbial taxonomy.

When Leeuwenhoek sent a seventeen-and-a-half-page letter to the Royal Society describing his discovery of animalcules for publication in their new, and first, scientific journal, *Philosophical Transactions*, it was met with such skepticism that even Hooke thought it was a delusion. Hooke sent an English vicar and some other reputable observers vetted by the Royal Society to Delft to verify the reports. The observers were as amazed as Hooke and his colleagues in London had been. In 1677, Leeuwenhoek's now verified observations were published by the Royal Society (in English, after being translated from the Dutch with help from Hooke, who had learned Dutch so that he could read Leeuwenhoek's papers). Leeuwenhoek was elected a Foreign Fellow of the society in 1780, but he never visited London.

Leeuwenhoek was a creative genius. He had no formal higher education and no affiliation with any university. He did not know Latin or Greek, the two languages of formally educated people at the time; he wrote only in Dutch. He built microscopes as a pastime and gave many of them away; he never sold any. He bequeathed twenty-six of his instruments to the Royal Society, all of which subsequently were "borrowed" by members of that esteemed group of scientists; all the originals have since disappeared. The rest of his collection was sold for the weight of the silver or other metals in the bodies of the instruments. Over his ninety-year lifetime, he sired five children. Only one, Maria, lived beyond childhood, and his scientific legacy almost died with his own death in 1723.

Although Leeuwenhoek is often viewed as the father of microbiology, Hooke was the collaborative agent who led him to fame. Like the relationship between Lyell and Darwin a century and a half later, Hooke and

Leeuwenhoek were symbionts. Both remarkable men were critical catalysts for the impending discovery of the invisible world. On a personal level, both were extremely generous toward each other to the end of their lives.

The descriptions and enumeration of microbes seemed to support the idea of spontaneous generation of life (in pepper infusions no less!), the idea that organisms could be formed from dead or nonbiological sources without any obvious parental lineage. For example, it was commonly accepted that maggots could form in dead meat and that wasps could come from buried elk horns. Spontaneous generation was widely believed by most people at the time. Leeuwenhoek rejected the basic notion, but he could not disprove it. The role of microbes in biological functions was virtually ignored, and it would be almost 200 years before these organisms would garner further serious attention. Amazingly, while the fundamental discoveries in science in the seventeenth century—gravity, light waves, planetary rotation around stars, and the incredible abstraction of science in mathematics—spurred huge explosions of discoveries in physics and chemistry, fundamental discoveries in biology largely lagged behind and were important only as they related to human health.

Neither Hooke nor Leeuwenhoek had students, and although *Micrographia* was a big seller in 1665 and for some years afterward, Leeuwenhoek never wrote a book, and his papers were not widely read. Neither Leeuwenhoek nor Hooke had a biological successor, and unlike Galileo, neither had immediate intellectual successors. Interest in pepper water faded. The microbial world was delegated to an invisible world in the eighteenth century—as natural philosophers turned to questions about the evolution of plants and animals and the sequences of geological structures that contained fossil remains of extinct organisms. One didn't need an expensive and delicate microscope to become an amateur scientist—all one needed was a hammer that could break rocks.

The renaissance of the study of microbes began only in the middle of the nineteenth century. It was championed by an almost forgotten hero, Ferdinand Julius Cohn. Cohn was a Jewish wunderkind who had been born in Breslau, Prussia (today's Wroclaw, Poland), in 1828. It is reported that Cohn learned to read before he was two years old, began high school at seven, and entered the University of Breslau at fourteen. Although he completed all the requirements for a degree, he did not receive one from the University of Breslau because of rampant anti-Semitism in Prussia at the time. He completed his studies at the University of Berlin, obtaining a doctorate in botany at the age of nineteen, and returned to the University of Breslau in 1849. In the same year, his father bought him the most expensive and best instrument available at that time—a microscope designed by Simon Plossl. I would have been very jealous of that microscope. Plossl was an Austrian optical-instrument maker who found a way to correct most of the optical aberrations inherent in microscopes and telescopes that contain several lenses. His lens designs are used to this day.

Cohn's interest in microbes was stimulated by his own observations with the gift of his father. At the University of Berlin, he was inspired to study single-celled algae by two remarkable professors: Johannes Müller and Christian Ehrenberg. The latter was one of the most famous scientists in Germany at the time. He had identified diatoms, a type of unicellular algae, in dust particles that Darwin had collected from the Azores during his trip on the Beagle. It was the first discovery that microbes could be transmitted long distances in the atmosphere by winds. Ehrenberg also showed that chalk was composed of fossils from microscopic organisms—and that observation would later become a model for looking for fossil microbes in rocks.

As Cohn's interests grew and the optics of microscopes improved, he became increasingly interested in algae and bacteria—or at least what he thought were bacteria. Having been traditionally trained in the biology of the time, he set out to classify bacteria in the context of other organisms. The classification of organisms in relation to others was a safe and obvious role for a biologist, and remains so to this day. Cohn never wrote about origins of life or the evolution of microbes, but he defined bacteria as unicellular organisms that lacked chlorophyll, the green pigment that characterizes algae and higher plants. Although he was well aware that most bacteria are not photosynthetic, Cohn classified them with algae, as plants. In the tradition of the time, Cohn attempted to organize microbes based primarily on their shape, a simple system that Leeuwenhoek had devised more than a century before and one that is still somewhat useful as a general guide (although it has been usurped by molecular sequencing technology in the twentieth century).

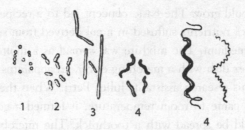

FIGURE 7. Illustration of the shapes of microbes described by Ferdinand Cohn in his treatise *Über Bakterien: Die Kleinsten Lebenden Wesen*, published in 1875. He characterized these organisms as related to algae and plants and assigned them to four families by shape: 1. the Spherobacteria (spherical bacteria); 2. the Microbacteria (short rods); 3. the Desmobacteria (straight filaments); and 4. the Spirobacteria (spiral filaments). This basic, simple system of descriptive classification was useful and persists to the present time.

Perhaps the most important of Cohn's contributions was that he rediscovered the field of microbiology. Like Leuwenhoek, he showed that microbes are all around us: in water, soil, and the air; in our mouths and guts; on our hands, clothes, and food. But, unlike most of his contemporaries, Cohn was not focused on the role of microbes in causing human disease. Although he worked on microbial diseases of plants and animals, and was far less famous than Pasteur, he had an even broader vision. He saw microbes as organisms that helped shape the chemistry of the Earth—the planet's metabolism. Early in my career, Cohn was an inspiration. He was an amazing pioneer of environmental microbiology.

One of Cohn's contributions to microbiology was the isolation of specific strains of microbes, that is, genetic variants of a species. He developed techniques to grow microbes in liquid media with a particular nutrient added to coax one strain or another to grow rapidly. In 1876, two hundred years after Leeuwenhoek had described microbes, a rural German physician, Robert Koch, visited Cohn to ask advice about the cause of anthrax. Koch had isolated a potential resting stage of the anthrax bacterium from soils and had developed a new technique for culturing it. His approach was simple, ingenious, and unique. It was based on isolating the microbes on the surface of a gel, where colonies, derived from a single cell, could grow. The basic concept led to a recipe that included the addition of nutrients suffused in a gel derived from seaweeds (agar) as a growth medium. The mixture was spread as a molten liquid on a small, flat, glass dish with a matching cover, an apparatus that had been designed by his research assistant, Julius Petri. When the medium with the nutrients came to room temperature, it formed a gel upon which microbes could be spread with a toothpick. The microbes would then form colonies and could be picked off the gel and regrown. This process could be repeated until only one strain of microbe was isolated. The use of agar and the dish for growing microbes made purification of the anthrax bacterium possible. It is amazing that Koch didn't become infected with his own cultures. Today we would be horrified if an amateur scientist cultivated anthrax in a lab in his or her home or garage.

Based on the purification techniques he developed with Petri, Koch described a set of postulates, which remain the foundation for identifying vectors of infectious disease to this day. They are as follows: (1) the microbe must be found in all cases of the sick organism and be absent in healthy ones; (2) the microbe must be isolated and maintained in pure culture; (3) the purified microbe must be, upon exposure, capable of infecting a healthy organism; and (4) the microbe must be identified and isolated from an exposed organism. By applying these four conditions, Koch experimentally proved that anthrax was responsible for the disease in cows, making it the first disease absolutely proven to be caused by a microbe.

Cohn was extremely impressed with Koch's logic and meticulous methods. He published Koch's paper in a botanical journal in 1886, and with Cohn's encouragement, Koch went on to show that cholera and tuberculosis were also microbial diseases. Koch received the Nobel Prize in 1905, and his postulates became dogma for decades. Koch's notion that microbes could be isolated and grown in culture pervaded the microbiological community for the first seventy years of the twentieth century. It was a logical idea, and it strongly influenced the identification of microbes in causes of diseases, but the dogmatic set of postulates also had the unintended consequence of setting back research in microbial ecology and evolution.

Over the decades, microbiologists patiently isolated species of microbes. There is no doubt that studying individual organisms in isolation has helped us to understand the basic features of how individual species make their living. But this approach also biased our understanding of how microbial communities function. It is akin to extrapolating the behavior of an African cichlid in my aquarium to their behavior in the lakes in which they live. An aquarium is not a natural environment. Neither is a Petri dish nor a test tube of liquid media containing nutrients thousands of times more concentrated than in the ocean or lakes. That scientists really did not know how to cultivate microbes would become apparent only in the latter half of the twentieth century, when they became aware that microbes are social organisms that live in complex communities. We will discuss the social organization of microbes a bit later.

In 1977, three hundred years after Leeuwenhoek reported the very existence of microbes, Carl Woese and his colleague, George Fox, both biochemists and geneticists at the University of Illinois, reported that all organisms in the world could be arranged into three major categories on the basis of one of their intracellular structures, the ribosome. It was well known that all microbes have ribosomes, but some organisms do not contain structures within their cells that are enclosed in membranes, while others do. The abstract of their paper, published in the *Proceedings of the National Academy of Sciences of the United States*, was a single sentence: "A phylogenetic analysis based upon ribosomal RNA sequence characteristics reveals that living systems represent one of three aboriginal lines of descent: (i) the eubacteria, comprising all typical bacteria; (ii) the archaeabacteria, containing methanogenic bacteria; and (iii) the urkaryotes, now represented in the cytoplasmic component of eukaryotic cells."

Even more important was the apparent relationship of organisms to each other. Not only are animals and plants just small twigs on the tree of life, but animals are most closely related to fungi. It isn't intuitively obvious that a mushroom is a closer ancestor to a mosquito or elephant or us than to a higher plant, but it is. Specifically, Woese and his colleagues demonstrated that all living organisms could be arranged on a tree of life based on the history of their protein-assembling machinery.

We all know some proteins—they are the stuff of egg whites, they are our skin, our hair, our finger nails, the meat in our muscles. They are the enzymes—the molecules that convert what we eat into energy and our bodies. Without proteins, cells would not be able to do any work. Without being able to work, a cell would not be able to replicate.

A key component in the formation of proteins are *ribosomes*. Ribosomes are complex nanomachines that are composed of both proteins and *ribonucleic acids*, or RNAs. Woese and Fox sequenced the RNA molecules in the ribosomes and discovered that there were subtle but consistent differences in the sequences within the twelve species of organisms they selected, including five species of bacteria, four species of microbes that produce the gas methane, a yeast, a small plant (duckweed), and a mouse cell. They found that the RNA sequences in ribosomes from bacteria were more similar to each other than they were to those of the yeast, plant, or mouse and were also distinctly different from those in the microbes that metabolize methane. This work demonstrated that although there are three superkingdoms of life, all living organisms are related to each other via the RNA sequences in their ribosomes.

Because all organisms have ribosomes, Woese and his colleagues postulated that all organisms on Earth are descendants from one single, but extinct, common ancestor. To imagine otherwise, one would need to invoke the most absurd and extravagant proposition, that is, that ribosomes evolved independently millions of times to create the spectrum of life forms we see today. In effect, Woese proved Darwin's idea that all life on Earth is connected to a common ancestor that arose long ago. The information in extant ribosomes potentially allows us to reconstruct relationships among all organisms. The basic evolution of the nanomachine that became the ribosome is obscure—but there could only have been one common ancestor from bacteria to us. That ancestor had to have been a microbe. Darwin, Hooke, and Leeuwenhoek would have been totally amazed that one could construct the relationship between all living organisms from the structure of a core machine that is responsible for the production of proteins.

In 1990, based on sequences of the nucleic acids in ribosomes that Carl Woese and his colleagues had worked on for several years, he constructed

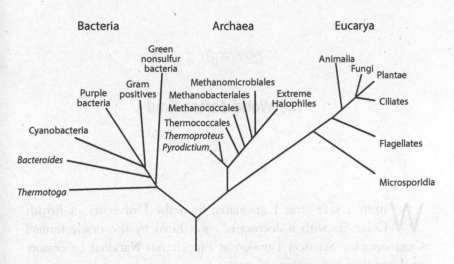

FIGURE 8. Carl Woese and George Fox's tree of life relates living organisms to each other based on ribosomal RNA sequences. Woese and Fox discovered that the bacteria are actually two super families of distinctly different organisms, Bacteria and the Archaea. Furthermore, animals and plants are subgroups within a larger family of eukaryotes, Eucarya. The vast majority of organisms in this tree of life are microbial.

a universal phylogenetic tree of life. The tree was fundamentally different from what Darwin had envisioned. Life on Earth is much more than plants and animals, and much, much more than Leeuwenhoek, Hooke, or even Darwin could possibly have imagined. The overwhelming majority of life on Earth is microbial! In fact, there are far more species of microbes than there are of all plants and animals combined. We don't really know how many there are, but the number is in the several millions, at a minimum. What we do know is that the basic structure of the tree of life has helped us understand that all extant life on Earth is derived from a single, extinct microbial organism.

But if all life on Earth is derived from a common microbial ancestor, when did that last common ancestor arise?

CHAPTER 3

||||||||

The World before Time

Within a year after I graduated from the University of British Columbia with a doctorate, I was hired by the newly formed Oceanographic Sciences Division at Brookhaven National Laboratory on Long Island. Brookhaven's major focus is on physics, and to some extent, chemistry. Although I was not a member of either the physics or chemistry departments, over the course of the next twenty-three years, I learned a lot from my physics and chemistry colleagues.

Physicists value simplicity. They try to strip natural phenomena down to the barest essentials. One of the intersections between physics and chemistry, a field called nuclear physics, became extremely useful in understanding geological processes. Early in the twentieth century, fundamental research in that area, specifically the discovery of isotopes by Harold Urey, a physical chemist, helped us to peer into a world before time.

A chemical element is defined by the number of positively charged particles, *protons*, in the nucleus of its atom. An isotope contains more or fewer neutrons relative to the number of protons. Neutrons have no charge, but they function as the "glue" in the nuclei of atoms, keeping the protons from repelling each other. Every element has several isotopes. For example, carbon contains six protons. The most abundant isotope of carbon contains six protons and six neutrons and therefore is denoted carbon-12. But there is also an isotope of carbon that contains six protons and seven neutrons (carbon-13), and one that contains six protons and eight neutrons (carbon-14). The former is stable—that is, it exists indefinitely; the latter is radioactive, that is, one of the neutrons decays to become a proton, thereby forming nitrogen-14, which is stable and exists indefinitely. When a neutron in carbon-14 decays to become a proton, the atom simultaneously emits a negatively charged particle, an *electron*, which is often called a *beta particle*. The emission of beta particles can be detected very accurately and so can be used for determining the abundance of carbon-14 in the original material. The half-life of carbon-14 is approximately 5700 years; that is, after about 5700 years one-half of the carbon-14 atoms in a population will have become nitrogen-14. The radioactive decay of carbon-14 potentially allows dating of materials that contain carbon, for example in bones, teeth, wood, and so on. But after tens of thousands years, virtually all the carbon-14 will have decayed, and the signal will be too weak to be useful for dating materials. Coal and petroleum, which were formed many millions of years ago, no longer have any detectable carbon-14; they are much older than several half-lives of the radioactive isotope. Fortunately, however, there are other naturally occurring radioactive isotopes with half-lives of hundreds of millions, even billions, of years. Two of these are isotopes of uranium: uranium-238 and uranium-235.

These two natural isotopes of uranium were formed in a very hot, very short-lived star that exploded, called a *supernova*, that gave rise to our solar system long before our star, the Sun, began to shine. The uranium isotopes were incorporated into meteorites as our solar system was formed. Uranium-238 has a half-life of 4.46 billion years, while that of uranium-235 is 704 million years. Ultimately these two isotopes decay to two different, stable (nonradioactive) isotopes of lead.

The study of uranium isotopes was strongly supported in U.S. national laboratories during the Second World War for the obvious reason that one of the isotopes could be used to make an atomic bomb. However, the discovery of uranium isotopes led to many practical applications aside from the production of weapons. Indeed, radioactivity in naturally occurring elements in rocks allows us to date events in Earth's early history, including the earliest evidence of microbial life.

In 1953, Clair Patterson, then a thirty-one-year-old chemist at the California Institute of Technology, measured the isotopes of lead in a meteorite found in Diablo Canyon, a crater that is in the northern part of Arizona and was formed about 50,000 years ago from the impact of a large meteorite. Because meteorites were produced during the early formation of our solar system, the age of a meteorite should roughly correspond to that of the formation of a solid surface on the Earth.

Patterson took samples of the meteorite to Argonne National Laboratory for analysis of the isotopes of lead, which he knew must have been derived from the decay of the two isotopes of uranium. Based on very careful analyses, he calculated the age of the Earth to be 4.55 billion years, a date that has withstood the test of further scientific scrutiny. The 300-million-year age calculated by Darwin nearly a century before Patterson's measurement of lead isotopes was off by more than a factor of 10!

What does the date inferred from lead isotopes mean? It means that this planet had formed a hard crust before 4.55 billion years ago. But if the Earth is so much older than Darwin could have ever imagined, when did life first evolve on the planet? The radioactive decay of uranium in meteorites like those Patterson studied is not sensitive to temperature— that is, the meteorite could have become very hot or very cold, and the calculated age would be exactly the same. But unlike meteorites, most of the rocks on Earth have undergone one or more episodes of change because the interior of the Earth is very hot. The heat is produced by radioactive decay of uranium and two other elements, thorium and potassium. In turn, the heat inside the planet produces volcanic eruptions and earthquakes on its surface. This process brings new materials to the Earth's surface but simultaneously forces sediments in the oceans into the interior of the planet, where they are melted. The further one goes back in time, the amount of rock from that time becomes increasingly smaller, the reason being that most of the very old rocks have been eroded to sediments, subducted beneath the surface, melted, and have formed new rocks. Although the process takes hundreds of millions of years, very few rocks escape from it; but even if some do and are not completely eroded, they are often subjected to changes in temperature and pressure that are large enough to destroy the remnants of any organic matter that may have been formed by life. Somewhat ironically, the elements that allow us to reconstruct the age of the Earth destroy the evidence of life in the oldest rocks on the surface of the planet.

There are a few places on Earth where very old rocks can be found that have not undergone extremes of heat or other episodes of change that alter their record of creation. The oldest such rocks are in southwest Greenland, in the Isua Formation, an area of the Earth that is one of the most interesting to visit. The rocks are all about 3.8 billion years old and are very easy to see because very little vegetation covers them. I spent a month there a few years ago with my friend and colleague, Minik Rosing, who has been studying rocks from this formation for decades. It is hard to see compelling evidence of life in them; there is no evidence of physical fossils. There is, however, a small vein of graphite in the Isua Formation. Graphite is a form of solid carbon and was a highly prized mineral in the sixteenth century because it was used to form molds for molten metal, for example, for cannonballs. While we may not know how cannonballs were made, we all know what graphite is: a powdered form of the mineral, mixed with clay that has been used to make the lead in pencils for the past two hundred years. In Isua, the graphite veins were derived billions of years ago from the heating of sedimentary rocks, rocks that originated in an ancient ocean.

The graphite from Isua is highly enriched in one of the two stable isotopes of carbon: carbon-12. This enrichment is curious because the primary cause of carbon-12 enrichment in organic matter is the result of photosynthetic processes. All photosynthetic organisms, such as the microbes I studied in the Black Sea, prefer to use the lighter, stable isotope of carbon to make their cells. Might the isotopic enrichment of carbon-12 in the graphite of Isua mean that there were photosynthetic microbes in the oceans 3.8 billion years ago? I am not sure we will ever know for sure, because the rocks from that area have been too altered by heat and pressure to infer much more from them; however, there are other, albeit younger, rocks elsewhere that have not been changed as much through time.

Two other major areas of ancient rocks are found in South Africa and Western Australia. The oldest rocks in these two regions date as far back as about 3.6 billion years ago, and some of them contain more concrete traces of life in the form of physical fossils and altered isotopic composition of carbon. One area where physical fossils are found is the

Strelley Pool Formation, in Western Australia, which contains evidence of microbes from about 3.4 billion years ago. Although it is very difficult to see and verify physical fossils of microbes, when any organism dies, there is an infinitely small chance that it leaves a biochemical trace in the sediments. In the case of microbes, the traces best preserved are usually from lipids—the fats that comprise membranes of the cells. These molecular fossils are found in rocks after the first 2.7 billion years of Earth's existence. It is very hard to find rocks much older that have not been heated or altered and so preserve any complex organic material at all. Unfortunately, neither ribosomes, or any other nucleic acids, nor proteins have been preserved in rocks over billions of years—if they had, our understanding of the history of life would be much more complete. In younger rocks, there is compelling evidence of microbial life. By approximately 2.6 billion years ago, there are clear, visible physical fossils of microbes in rocks and the variations in the isotopes of carbon, nitrogen, and sulfur present strong evidence of a rich microbial world in the oceans of that time.

Based on both molecular (mostly lipid-derived molecules) fossils as well as physical fossils, an interpretation of the rock record is that during the first 3.5 billion years of Earth's history, or for about 85% of the time since the planet was formed, life was completely microbial and almost entirely restricted to the oceans. There were no animals, no land plants, no true soils, and for a very, very long time, virtually no oxygen.

But can we say anything about how these ancient microbes actually functioned at that time? And can this tell us anything about the rise of plants and animals 3 billion years later?

An analogue of the ancient microbial rock record is the Black Sea. Indeed, in many ways, the deep water of the modern Black Sea appears to harbor many similar types of organisms as might have been found in the oceans about 3 billion years ago.

Why do we think that Black Sea is a contemporary analog for a lost microbial world?

In 1997, Bill Ryan and Walter Pitman from Columbia University suggested that about 7500 years ago, as the ice sheets in the northern hemisphere melted, water from the Mediterranean flowed through the

Bosporus Strait and flooded the Black Sea. Their hypothesis was that the flood was rapid and potentially was the true basis of the story of Noah and the ark. Regardless of whether the Black Sea flooded suddenly or more gradually, as others contend, the result is that warm, very salty water entered the basin through the narrow, shallow sill that separates the European side from Asian side of modern Turkey. This salty water is denser than the freshwater that flows into the basin from the Don, the Dnieper, the Danube, and other rivers to the north. The denser salty water sinks into the deep basin while the overlying water is relatively light. The differences in the physical density of the water masses make it virtually impossible for the deep water to come to the surface, where it can be oxygenated by the atmosphere. Consequently, as organic matter, which is produced by photosynthetic organisms in the surface, sinks into the interior of the Black Sea, it is consumed and respired by microbes, depleting all the oxygen in the interior of the Black Sea. Indeed, the interior of the Black Sea has been anoxic for thousands of years. It is the only semi-enclosed basin that has been anoxic for so long. How do we know this?

As a result of nuclear weapons testing in the 1950s and '60s, a large quantity of carbon-14 was produced and spread across the atmosphere. Some of that carbon came into contact with the surface water of the oceans, and as the water from the surface was carried into the interior of the oceans and seas, the radioactive decay of the isotope could be precisely measured and followed, providing a kind of clock. By calculating back to the initial concentration of carbon-14 in the atmosphere, oceanographers could determine how long ago the water in any ocean basin was exposed to the atmosphere. Based on such an analysis, the deep water of the modern Black Sea was last exposed to the atmosphere about 1500 years ago, and while that is not a long time from a geological perspective, it has been long enough for any oxygen produced below the upper hundred meters to have been consumed very rapidly, once the water sank again. The interior of the modern Black Sea has been anoxic for at least the last 8000 years.

Although the microbes in the deep Black Sea are not literally billions of years old, they are living fossils in that they retain metabolic processes—or, simply the internal machinery—that evolved very early in Earth's history. In effect, they preserve the metabolism of organisms that pervaded the world's oceans billions of years ago. By understanding their metabolism, we can get a sense of how life worked in a world long since and forever gone. But we can do more than understand how life worked billions of years ago. By studying this ancient microbial machinery, we can also understand the connections between microbes and all other plants and animals, including ourselves.

Let us take a look "under the hood" to see how some of the machinery that makes these invisible creatures work. Let's explore how microbes developed the machinery within their cells that would become the engines of life on Earth and the key to planetary habitability.

CHAPTER 4

॥॥॥॥॥

Life's Little Engines

Little could Robert Hooke foresee the significance of his description of
microscopic cells in the thin section of cork he cut with a penknife.
Over the course of the more than three centuries since Hooke first
described the outline of their structures, scientists have spent much time
and effort to understand how cells—the smallest form of life capable of
self-replication—work. Most of that effort has led to a quest to understand the machinery inside cells that allows them to obtain energy, grow,
and reproduce. While we don't know all the answers, we do know that,
like matryoshka dolls, within the discrete containers of cells themselves,
there are smaller containers that carry out specific functions. For want
of a simple term, I call these smaller containers within cells life's nanomachines. They are assemblies composed largely of proteins and nucleic
acids and carry out the necessary functions of all living cells. I have spent
a good deal of my scientific life trying to understand how they work.

Understanding how these nanomachines function is important because
their internal workings allow us to see how basic processes were copied
and repackaged in different forms. The concept is not that far from taking parts from an electronics supply store and building amplifiers, radios,
televisions, or whatever device can be made. Nature boasts many different types of nanomachines. As I discussed earlier, one of the oldest—the
ribosome—evolved in a microbial ancestor billions of years ago. We will
return to that ancient world of early microbes in chapter 5, but first let's
see about other nanomachines and how they actually work within cells.

In some ways, trying to study the machinery inside living cells is
analogous to trying to understand how a car works with no knowledge

of what is under the hood. We see cars traveling around on the streets, and it is clear that they have some mechanism that allows them to move. We can stop the car and take the key out of the ignition, and the car won't go. If we can open the car's hood, we can potentially dissect the machinery and examine all the parts—down to every bolt, every washer, every gasket. And if we take an even closer look, we can see that the parts are assembled in a very precise way, but there are no instructions as to how to assemble them. Unless we understand what the parts do, we can't imagine how the machinery allows the car to travel down the road. But pulling out a piston or a battery, let alone a computer, can potentially give clues as to what that specific part does and how it functions in the machine.

Using the analogy of trying to understand how a car works for learning how cells function is obviously imperfect. Cells are a lot more complicated than cars. Cars don't assemble themselves, they don't replicate themselves, and, unfortunately, they don't repair themselves. It probably isn't too surprising, then, that biologists have taken parts out of cells to see how the individual components work but thus far have failed to reassemble the parts of a cell from scratch into a fully functional, self-replicating organism. We have a long way to go in understanding what is "under the hood" of cells. However, during the three hundred years since Hooke described the basic structure of cells, we have made a lot of progress in identifying many of the key parts, and we are starting to see how the nanomachines inside cells work. That knowledge has allowed us to see patterns of organization of cells across the tree of life. Indeed, it has given us an opportunity to understand what life actually is. But before we get into the nuts and bolts, so to speak, let's take a brief look at how the parts were identified.

The identification of the parts began in the nineteenth century, with the improvement of microscopes and the inquisitive and patient nature of mostly wealthy, male biologists. In 1831, a Scottish botanist, Robert Brown, distinguished by careful microscopic examination an opaque spot in the center of orchid cells, and later in pollen. In a paper presented to the Linnean Society in London, he called the structure a *nucleus*; it was the first intracellular structure identified. In 1869, a Swiss doctor working in Germany, Friedrich Miescher, discovered that the intracellular structures identified by Brown contained interesting molecules that were not proteins, and he called the new material *nuclein*. Almost a century later, those molecules would be found to carry the information to make new cells.

In the latter part of the nineteenth and early twentieth centuries, lens makers developed increasingly better lenses and other optical components for light microscopes that allowed one literally to peer into large cells. The visualizations were made even better with stains and dyes that bound to specific components. These types of advances led to a very basic understanding of the arrangements of some of the components in eukaryotic cells, that is, cells that contain nuclei. Plants and animals are essentially organized conglomerations of eukaryotic cells.

With better lenses, stains, and microscopes that had even higher magnification, several discoveries occurred in a relatively short time. In 1883, another botanist, a German, Andreas Schimper, discovered that starch, which stains dark brown in the presence of iodine, was made in the miniature green bodies of plants, which he called *chloroplasts*. In 1890, yet another German, Richard Altmann, recognized that a population of small particles appeared to be present in all animal cells. He called them *bioblasts*; they would later come to be known as *mitochondria* (singular, *mitochondrion*). Altmann also found that Miescher's nuclein was acidic and renamed the substance *nucleic acid*. In 1897, an Italian physician, Camillo Golgi, described yet another structure, which would come to be called the *Golgi apparatus*. At first it was thought that the structure was an artifact of the stains Golgi used, and it was not until the middle of the twentieth century that it was confirmed as being a real entity. Several other large structures would be described later by very patient observers working with the best optical microscopes produced at the time. But regardless of how good lenses are, there is a physical limitation of what can be seen with a microscope that uses visible light.

Structures that are smaller than about 1000th of a millimeter (i.e., a micrometer) are simply very hard to see in detail with visible light. A human hair is about 100 micrometers in diameter. The diameter of most bacteria and other microbes is about 1 to 2 micrometers; sometimes even less. To put that into perspective of our naked eye, one could line up about 100 of these cells to span the diameter of a human hair. And because microbes are so small, it is virtually impossible to discern the structures inside them. Were there miniature nuclei? Mitochondria? Chloroplasts? This quest to understand the small structures inside cells was reminiscent of Leeuwenhoek's earlier concept of visualizing animalcules as miniature animals. For several decades, scientific progress in distinguishing very small cells or small parts within large cells was at a standstill, limited by the resolution and magnification of light microscopes.

A big breakthrough came in the 1930s, when two German physicists—Max Knoll and his student Ernst Ruska—developed a new type of microscope that used high-energy electrons that were accelerated as beams through a vacuum and onto a sample, which either scattered, absorbed, or transmitted the electrons. The resulting image could resolve structures down to tenths of micrometers, over a hundredfold higher magnification than could ever be achieved with a light microscope. A whole new world opened up—one in which, for the first time, we really could "look under the hood" into cells.

Examination of cells with the electron microscope quickly confirmed the existence of nuclei, the Golgi apparatus, mitochondria, and chloroplasts in eukaryotic cells. But surprisingly, it also revealed that these structures were absent in many microbes. There appeared to be a finite number of matryoshka dolls in microbes. The organisms that lacked these internal, membrane-bound structures were collectively lumped into a group called *prokaryotes*. However, details about the architecture of the interiors of all cells revealed some common structures, regardless of whether the cell had a nucleus or not. All required certain parts.

One of these all-inclusive parts is the ribosome. They were first discovered in 1955 by a Romanian biologist, George Palade, working at the Rockefeller Institute (now Rockefeller University), in New York. Using the best electron microscopes available at the time, Palade described the structures in images from mammal and bird cells—both eukaryotes. Ribosomes looked like very small, fuzzy balls that appeared to both float freely in the liquid inside cells and line up along specific internal membranes.

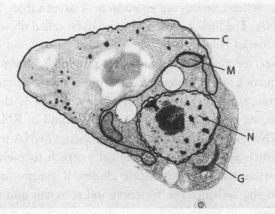

FIGURE 9. An electron micrograph of a thin section of a green algal cell. This organism is a eukaryote (see Fig. 8), and like all eukaryotes, contains several internal organelles that are bound by membranes. In this algal cell, the organelles include a chloroplast (C), mitochondria (M), a nucleus (N), and a Golgi apparatus (G). (Original photomicrograph by Myron Ledbetter and Paul Falkowski)

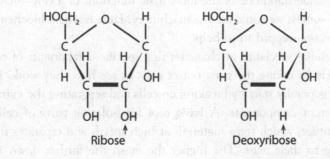

FIGURE 10. A diagram of the structures of ribose and deoxyribose. The former is found in ribonucleic acids (RNA) the latter is in deoxyribonucleic acid (DNA).

Palade discovered that the small balls contained both protein and a nucleic acid, but the function of the little components would not be understood for more than another decade. It was clear, however, that the nucleic acid in the nucleus was DNA, whereas that in the ribosome contained ribonucleic acid—another type of nucleic acid with a different sugar, ribose, which contains one more atom of oxygen than deoxyribose, found in DNA. The little balls would come to be called *ribosomes*, which is a contraction of "ribose" and "some" (body).

Ribosomes are microscopic machines that take information from a DNA sequence via a messenger molecule. The messenger is a mirrored, or complementary, pattern of a gene, which is the template of a protein sequence. The complementary strand of RNA is called *messenger RNA*. The information in the messenger RNA instructs the ribosome which amino acids to chemically attach to each other and in exactly what order. The resulting chains of amino acids become the proteins the cells need to function and to repair and make more of themselves.

Because all the basic building blocks of cells either are proteins or depend on proteins for their formation, ribosomes are absolutely essential components in every cell, but they are very complicated machines. They are only about 20 to 25 nanometers in diameter (a *nanometer* is 1000th of a micrometer, which is 1000th of a millimeter)—which makes them hard to see, even with an electron microscope. It was a dilemma: how could we understand one of the most basic functions of a cell—making proteins—unless we could see the machinery? Here is where biochemists and physicists stepped in to help.

Biochemists specialize in characterizing specific components of cells, particularly by taking the parts out of cells to see how they work. Biochemists generally start by breaking up cells and separating the extracts into different components. A basic tool for isolating parts of cells is the centrifuge, which spins materials at high speeds and separates them according to their mass. The higher the mass, the further down in a centrifuge tube the material or particles will go. Using a very high speed centrifuge, Palade isolated the fuzzy round ball structures that he had seen in the electron microscope.

But the question remained: how do ribosomes actually work? By isolating ribosomes, Palade and others determined that the structures contained proteins and still another kind of RNA molecule that was different from that in the messenger RNA. It was soon shown that these tiny balls could form proteins in a test tube if one provided the proper components.

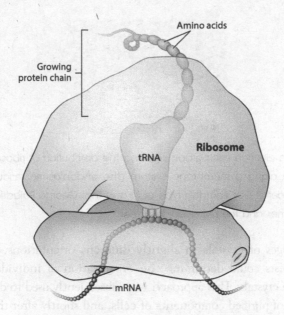

FIGURE 11. A cartoon showing the basic function of a ribosome. This nanomachine makes proteins using a template of information originally encoded in DNA and transcribed by a messenger RNA molecule. The messenger RNA molecule provides the information for the sequence of amino acids for a specific protein; each protein in a cell has a specific messenger RNA. The ribosome, which also contains RNA but is organized into a larger structure with many proteins, "reads" the information from the messenger RNA and uses a third RNA molecule with a specific amino acid attached (transfer RNA) to build up the protein one amino acid at a time. The protein emerges from the ribosome to finds its proper place within the cell.

But even the best electron microscopes couldn't resolve what was inside the ribosomes that Palade had isolated. Solving that problem needed an even more powerful imaging tool.

In the early twentieth century, shortly after the discovery of radioactivity, it was realized by physicists that X-rays, which are very high energy particles of light, are scattered by crystals in a very organized fashion. X-rays have much more energy than electrons and can resolve very tiny structures—even down to the level of individual atoms. By taking many

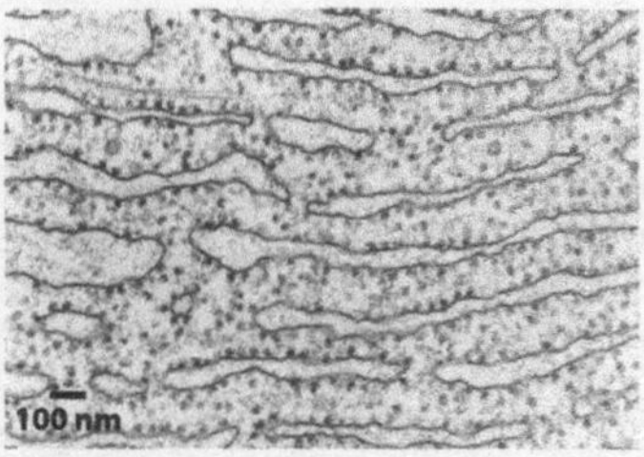

FIGURE 12. An electron micrograph showing the distribution of ribosomes (little fuzzy balls) along a membrane system (the endoplasmic reticulum) in a eukaryotic cell. It was from this type of image that George Palade first identified ribosomes and then later isolated them.

X-ray images of crystals in slightly different orientations, physicists and chemists could determine the organization of individual atoms within the crystals. This approach was subsequently used to discern the structures of purified components of cells, and shortly after the Second World War, determining the organization of atoms in crystals of proteins became feasible. It was very tedious work; hundreds of X-ray images had to be obtained and overlaid—without the aid of computers. By back calculating the angle of scattering of the X-rays from the crystal, physicists and chemists could deduce the structure of the molecule, even though it could not be seen directly with a microscope. As computers and very high energy X-ray sources, such as synchrotron light sources, one of which was across the street from my building at Brookhaven National Lab, became increasingly available, more and more structures of proteins were determined. Protein structures are archived in the chemistry department at my university, and anyone with a computer can see them online.

Ribosomes are not made only of a single protein, and they aren't simply proteins; they are much more complex structures. The simplest ribosomes, which are found in prokaryotes, contain not only RNA molecules but also about 60 proteins organized in two units. It was thought to be foolhardy to try to crystallize an intact ribosome, let alone obtain any useful information about their structure from X-rays. However, in the late 1980s, two scientists did. One, Harry Noller, was an American; the other was Ada Yonath, an Israeli biochemist working in Germany and Israel. With a lot of patience, perseverance, and insight, they produced the first X-ray images of ribosomes.

During the next two decades, several groups around the world began to analyze the structures of these amazing nanomachines. From very careful analyses of many X-ray images, Noller at the University of California at Santa Cruz, Yonath at the Weizmann Institute, Thomas Seitz at Yale University, and Venkatraman (Venki) Ramakrishnan, who was at Brookhaven National Lab (and a colleague) and then went to Cambridge University, pieced together how the ribosome works. The latter three shared the 2009 Nobel Prize in Chemistry for their efforts.

The two major complexes of the ribosome interact something akin to how a pair of gears works. The amino acids are ferried to the ribosome by a third RNA molecule, called *transfer RNA*. As the messenger RNA is fed into the ribosome like a piece of spaghetti, the two protein complexes move back and forth, attaching the appropriate amino acid to the previous amino acid to form a protein. This protein factory thus "stamps" out the information in the genes. This intricate machine works amazingly efficiently—between 10 and 20 amino acids per second are added to the emerging string of protein.

This complex protein factory is virtually identical in every living cell. There are small variations in the RNA within the ribosomes, but those variations are assumed to be *neutral mutations*, which occur all the time in nature. They are accidents that occur randomly and do not affect the outcome of the process. We can see neutral mutations all around us. Each of us has slightly different fingerprint patterns. Some of us have whorls, others have arches or ridges or loops. There is no correlation between our tactile sensitivity and our fingerprint pattern. Similarly, the mutations in ribosomal RNA do not appear to affect the rate at which the ribosome makes proteins. There are no "supersomes" and "wimpysomes" (at least we don't think so). In fact, the structures of all ribosomes are so similar that they barely can be distinguished; however, there are small differences between the nucleic acid sequences in the RNA within ribosomes. Those differences allowed Carl Woese and George Fox to separate the prokaryotes into two different supergroups—the bacteria and the archaea, which were in turn very different from the eukaryotes. While the differences in the nucleic acid sequences in the ribosomal RNA allows us to trace the evolutionary history of all living organisms, the differences in the RNA sequences do not affect the basic function of the ribosome. All cells make proteins in exactly the same way.

But making proteins isn't simple. Amino acids don't spontaneously form chemical bonds with each other. To form the bonds between two amino acids requires energy. So where does the energy to make the proteins come from? It is made by another set of nanomachines, which are found elsewhere in the cell. Here the world inside the cell gets even more bizarre.

The basic currency of energy in all cells is a molecule called *adenosine triphosphate* (ATP), a single nucleic acid molecule that is found in both DNA and RNA and contains a sugar and three phosphate groups linked one after the other. When this molecule is used in a biochemical reaction, it is cleaved to *adenosine diphosphate* (ADP) and a lone phosphate. The cleavage of ATP produces chemical energy, which is used for many purposes. One of the major functions of ATP in all organisms, especially in microbes, is in the synthesis of proteins. Another is for motility. Yet another is to pump ions, such as protons, sodium, potassium, and chloride, across membranes. All these functions and more are found across the tree of life. The universal distribution of ATP in all cells on Earth raises the question, how do cells make ATP?

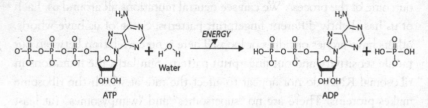

FIGURE 13. The basic currency of biological energy across the tree of life is adenosine triphosphate (ATP). When ATP is combined with water in enzymes, a phosphate group can be cleaved from the molecule to form adenosine diphosphate (ADP) and inorganic phosphate. That reaction releases the energy that all cells use for life.

The discovery of how most ATP is made in cells was extremely contentious, yet one of the most profound in biology. For many years, ever since Pasteur's discovery that microbes can use glucose as an energy source under anaerobic conditions, it had been known that ATP could be made in cells by transferring the phosphate group of some small molecules directly to ADP to form ATP. For a long time, that process, called *substrate phosphorylation*, was thought to be the only source of ATP, but the numbers just didn't add up. While in the absence of oxygen the amount of ATP produced by microbes was often low, in the presence of oxygen far more ATP was produced than could be accounted for by substrate phosphorylation. There had to another source of ATP.

In the 1950s, a somewhat eccentric English biochemist, Peter Mitchell, then working at Cambridge University, was thinking about how ions are transported across membranes. Membranes act as barriers to the diffusion of soluble atoms or molecules that carry an electrical charge, which are known as *ions*. He knew that in microbes, ATP could be used to transport ions and other molecules into and out of cells, across their membranes. But one of his graduate students showed that in a bacterium, the flow of sugars into the cell was accompanied by the flow of hydrogen ions (protons) out of the cell. The flow of the sugars and protons was dependent on ATP. Mitchell thought that if the reaction worked in one direction, it might work in the opposite direction—that is, by adding protons to a cell, it might make ATP rather than consume it. He left Cambridge and worked out of a laboratory on a small estate he had renovated in Cornwall. There he came up with a novel idea.

It was known not only that the structure described by Altmann seventy years earlier, the mitochondrion, is responsible for the production of large quantities of ATP but also that the rate of production of ATP depended on the presence of oxygen. The oxygen was converted to water—meaning that two atoms of hydrogen (H) were added to each atom of oxygen to make water (H_2O).

Mitchell proposed that there was a force across the membranes inside the mitochondria related to the concentration of protons in the organelle. He had discovered that inside the mitochondrion there was a network of membranes and there were more protons on one side of the membranes

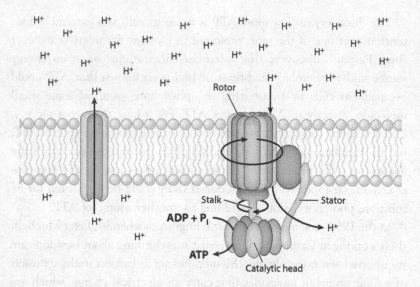

FIGURE 14. Adenosine triphosphate is made in cells by generating gradients of electrical charge across a membrane. In many cells and in two organelles, the mitochondrion and the chloroplast, the charge gradient is created by a proton gradient—that is, more protons (hydrogen ions) on one side of a membrane than the other. As the protons are funneled through a coupling factor embedded within the membrane, ATP can be made (see Fig. 15).

than the other. When the protons moved from the side where they were more concentrated to the side where they were less so, ATP was formed. The process, which Mitchell called *chemiosmosis*, required that the membranes inside the mitochondria remain intact.

Shortly after Mitchell published his hypothesis in 1961, a young researcher at Cornell University, André Jagendorf showed that a similar process exists in chloroplasts. Jagendorf isolated chloroplasts from leaves and bathed the organelles in an acidic solution but kept them in the dark. The chloroplasts could not photosynthesize because there was no light, but the inside of the organelles became acidic. He then transferred the chloroplasts to a neutral solution in the dark and showed that as the protons flowed out of them, ATP was formed. It would take another two decades to reveal the responsible machinery and how it operated, but in

1978 Mitchell won the Nobel Prize for his discovery of the chemiosmotic process of energy production.

The fundamental phenomenon that Mitchell revealed is that life uses electrical gradients to generate energy and that it uses energy to generate electrical gradients. The process is analogous to the way a battery operates. In effect, all organisms are electricity-generating systems—they work on moving the ions, such as protons, across a membrane and generate their own electrical gradients. The source of the protons and electrons is hydrogen—the most abundant element in the universe. The electrical gradient requires a membrane, without which there would be no difference in concentration in protons or other ions, and therefore no energy source to make ATP. Mitchell's discovery helped pave the way for understanding how the structures responsible for the ATP production work. These nanomachines are called *coupling factors*.

Coupling factors are literally miniature motors that span membranes. They contain a shaft that is a set of proteins spanning the membrane and physically inserted into a set of larger proteins (a *head group*) that sits at one end of the shaft. The basic design is something like a micro merry-go-round. Protons on one side of the membrane bind to and move along the shaft to cross the membrane. In the process, their flow physically turns the shaft counterclockwise, somewhat like how a waterwheel turns as the water flows across it. As the shaft physically turns, it mechanically moves the larger proteins (the deck of the merry-go-round), which bind ADP and phosphate. The deck oscillates, and approximately every 120 degrees of turn in the shaft, a molecule of ATP is formed and released to the cell for use in other functions. The motor can also operate in reverse. If there is a lot of ATP in the cell, it can pump protons, or other ions, across the membrane, and the ATP is converted into ADP and a lone phosphate.

This basic design of the miniature electrical motor for the production of ATP is very ancient. It evolved in microbes so long ago that we have difficulty understanding its evolutionary history. It is found everywhere in nature. In all animals it is a critical component of muscles and nerves. It is found in the roots and leaves of plants. It is found in microbes. Because the production of ATP is so critical to all organisms and so

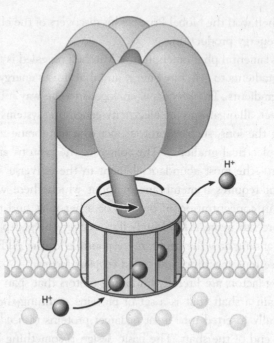

Figure 15. A cartoon showing the basic mechanism by which a coupling factor generates ATP from the flow of protons. The protons pass through a stalk in the membrane; as they do so, the stalk physically turns, and the head of the nanomachine, which is on the opposite side of the membrane, oscillates. The physical oscillation allows ADP and inorganic phosphate (see Fig. 13) to attach to the head group, where they are chemically bonded to form ATP.

dependent on membranes, all organisms must maintain an electrical gradient across their cell membranes. Among other things, electrical gradients are essential for transporting essential nutrients into cells and for transporting waste products out. But the electrical gradients produced by operating a coupling factor in "reverse" consumes energy.

Somehow, somewhere, biological machines have to acquire energy from the environment to generate the intracellular energy required to create the electrical gradients, or otherwise life would rapidly stop. The energy that powers all life on Earth is ultimately derived from the Sun.

Photosynthesis led to the evolution of the most complex biological reactions in nature. I have devoted most of my career to understanding how the process works. The core of the process occurs in yet another set of nanomachines, found only in photosynthetic organisms.

In photosynthetic eukaryotic cells, such as algae and higher plants, the responsible nanomachines are found only in the chloroplasts. However, the basic design of the photosynthetic process was first discovered in bacteria that do not split water but instead utilize molecular hydrogen. Regardless of what the substrate for the photosynthetic process is, the nanomachines responsible for converting light energy into chemical energy are called *reaction centers*. Like the coupling factor, they consist of groups of proteins that are embedded in membranes. The groups of proteins hold pigments, such as chlorophyll, and other molecules in specific positions so that the photobiological reaction will work. The proteins are, in the parlance of biochemists, "scaffolds" for the working parts of the nanomachine.

The photosynthetic process is almost magical. Light is absorbed and a chemical bond is made. What has the magical nanomachine done to convert the energy in the individual particles of light (*photons*) to a sugar—the stuff we, and virtually any self-respecting microbe, will use as a source of energy?

In photosynthesis, light is absorbed by a specific molecule, most commonly the green pigment, chlorophyll. The absorption of light at specific wavelengths, or colors, by specific chlorophyll molecules leads to a chemical reaction. When one, very specific chlorophyll molecule embedded in a reaction center absorbs the energy from a photon, the energy of the light particle can push an electron off the chlorophyll molecule. For about a billionth of a second, the chlorophyll molecule becomes positively charged. (You may recall those nerdy T-shirts with one stick figure talking to another. The first one says, "I lost an electron." The second asks, "Are you sure?" The first replies, "I'm positive.")

There is no such thing in a cell as a free electron. Once an electron is removed from a molecule, it has to go somewhere. One possibility is that it returns to the molecule from whence it came—and that does happen once in a while—but rarely. However, when it does happen, the reaction

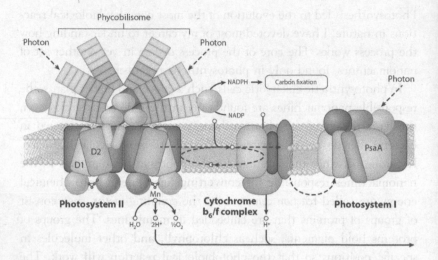

FIGURE 16. A schematic illustration of a reaction center in oxygen-evolving organisms. This is the only biological nanomachine capable of splitting water. It is composed of many proteins, and its primary role is to use the energy of the Sun to split water into oxygen, hydrogen ions, and electrons. The structure is embedded within a membrane, and the hydrogen ions from the water-splitting reaction are deposited on one side of the membrane. They flow through the coupling factor (Fig. 15) to generate ATP and eventually meet up with the electron on the other side of the membrane.

center emits red light—it literally glows. But more frequently, the energy of light is sufficient to push the electron on to another molecule that really doesn't need it but will temporarily accept it. How does that work?

Let's imagine for a moment you are an electron waiting for subway during rush hour. A train arrives at the station, but it is fairly full of other electrons. Now it is obvious that as an electron with a negative charge, you don't want to be in the same car packed with a whole lot of other electrons, each of which has a negative charge. The atmosphere in the train is very negative. But as the doors open, a man dressed in a uniform and white gloves pushes you into the car. (This really happens during rush hour in some cities.) The man in the uniform acts like particle of light—pushing you into an environment where you don't want to be—an environment with many other electrons. All the electrons crammed into this train make the car very negatively charged, but as the train moves along to other stations, electrons start to jump off, attracted to places where there are fewer electrons. As that happens, they start going to work, looking for places that are more positive. A similar thing happens on a very microscopic scale in reaction centers. But something even more interesting occurs.

The electron that was originally pushed out of the chlorophyll molecule in the reaction center by the particle of light left a "hole," and the chlorophyll molecule is now positively charged. To fill the hole, the chlorophyll molecule takes an electron from nearby molecules. In the case of oxygen-evolving organisms, such as blue-green algae, eukaryotic algae, and all higher plants, the electrons come from a quartet of manganese atoms held in a special arrangement on one side of a membrane. After donating their electrons to chlorophyll, the manganese atoms also need to fill their electron holes. They find water right nearby, and one by one, they extract four electrons from two water molecules using the energy of four pushes from photons, one at a time. As the water loses its electrons, protons fall off, and ultimately oxygen is left on its own to search for electrons. Oxygen is very keen to find electrons in nature, and that is why we term a molecule that wants to extract electrons from another molecule an *oxidant*. In other types of photosynthetic reaction centers, the electron source may be that rotten-egg-smelling gas, hydrogen sulfide, or in others it is a form of iron ions; in some others, it is carbohydrates (CH_2O). Regardless, the result is that ultimately all sources of electrons are external to the organism, and the primary use for all the electrons is to make sugars.

Whatever the source, invariably the electron is sent along one path while the proton sent along another. The proton, being positively charged, can be used to do some work as well. It is first deposited on one side of a membrane. The membrane prevents the proton from simply going to the other side, and the result is that on one side of the membrane there are many positively charged protons relative to the other side. This is, in effect, a miniature battery that can be used to generate ATP. But how can the protons do double duty: how do they recombine with electrons to make hydrogen, the element that is needed to make organic matter? Let's see how that microscopic contraption works.

Recall that reaction centers sit in membranes and that membranes are barriers to the free movement of protons and other charged molecules. As electrons are extracted from water or hydrogen sulfide, protons are formed. The protons concentrate on one side of the membrane. The membrane is a continuous sheet, something like a pita bread with the protons trapped in the pocket. After a few minutes of working in sunlight, the photosynthetic reaction centers can deposit 1000 times more protons in the pocket than are found in the environment outside; this results in 1000 times more positive charges on one side of the membrane than the other. The protons pass to the other side of the membrane through the coupling factor machinery, turning the motor and making ATP. That process occurs in every photosynthetic organism and is the major biological source of electrical energy in nature.

But what happens to the protons as they pass through the coupling factor and get to the other side of the membrane? They meet electrons and bind to another modified nucleic acid. That molecule has the unfortunate name of *nicotinamide adenine dinucleotide phosphate*, or NADP. When the proton and electron are added to NADP, the molecule becomes reduced, to NADPH. The role of NADPH is to ferry hydrogen around a cell so that it can be used to make organic matter. It seems like an unnecessarily complicated process, but if a cell were to make free hydrogen, the gas would be so physically small it could easily escape from the cell. By separating the two components of hydrogen, the electron and the proton and then reuniting them on a big molecule like NADP, the cell can trap hydrogen. In photosynthetic organisms, the hydrogen atoms on NADPH ultimately are used to convert carbon dioxide (CO_2) to sugars, which most of the rest of life on this planet uses to make its energy.

It took a lot of patience and some luck, but the crystal structure of a reaction center in a photosynthetic bacterium that does not split water was analyzed by three German biochemists: Hartmut Michel, Johann Deisenhofer, and Robert Huber. Their results, published in 1985 in the British journal *Nature* clearly showed how a core of three proteins in the heart of the reaction center held a bacterial chlorophyll and other molecules to form the working nanomachine. They won the Nobel Prize for Chemistry in 1988. Several years later, the crystal structures of the reaction center that split water was also elucidated, first by another German group, and later by several others around the world. We can see the parts of the machine, but unfortunately, we don't really see them working—yet. Rather than being movies of the machinery, X-ray analyses are snapshots. They catch the machinery in one specific state and do not reveal the motions of the machinery as it functions. While that deficiency has hindered a perfect understanding of how reaction centers actually work, we have come a long way toward understanding how light energy is used to split water and make oxygen.

Reaction centers are special: when they work, the nanomachine becomes a literal microscopic sound and light show. Recall that the energy of light pushes the electron from a chlorophyll molecule on the donor side of the protein complex to the acceptor side. The result is that for a billionth of a second there is a positively charged molecule and a negatively charged molecule inside a protein scaffold, and they are separated by only a billionth of a meter. Positive charges attract negative charges. The protein scaffold actually collapses slightly due to the attraction of the charges, and when it does, a pressure wave is created. The pressure wave is analogous to two hands clapping. Every time reaction centers move electrons, they make a microscopic clap, a sound that literally can be detected by a very sensitive microphone. This phenomenon, called the *photoacoustic effect*, was discovered by Alexander Graham Bell, the inventor of the telephone. In 1880 he used the effect to generate sound waves from light and made a device, the photophone, to transmit the sound. Who knew that the phenomenon could be used to listen to the sound of the engines of photosynthetic organisms as they pound out electrons? With my long-time colleagues and friends—David Mauzerall, at Rockefeller University; Zvy Dubinsky, at Bar Ilan University; and Maxim Gorbunov, in my laboratory—we have developed an instrument to measure the sound of the photosynthetic apparatus in living cells. Our analysis of the sound reveals that approximately 50% of the energy of light is converted to electrical energy in reaction centers.

But there is another signal that shows how photosynthetic reaction centers work. The reaction centers also change their fluorescence properties. When exposed to blue light, chlorophyll glows red in the process of *fluorescence*. We see fluorescence in DayGlo paints and in our teeth and on some cool T-shirts when we are exposed to ultraviolet light. But in photosynthetic organisms, the intensity of the fluorescent red light increases when more and more reaction centers are working. Briefly, when algae or leaves are in the dark and exposed to a blue light, the intensity of emitted red fluorescent light rises rapidly. The phenomenon was first reported in 1931 by two German chemists, Hans Kautsky and A. Hirsh, who observed the effect with their naked eyes. Over the next seventy years, the phenomenon was shown to be a quantitative measure of how much work reaction centers do. Consequently, it is now measured routinely throughout the world with sophisticated instruments to study how much sunlight is converted into useful energy in photosynthetic organisms. I have also spent many years of my research career using that approach to understand photosynthetic energy conversion efficiency across the world's oceans. Indeed, these types of instruments, which are capable of detecting fluorescence, were what I brought to the Black Sea to look for photosynthetic reactions in the oceans.

There are many other nanomachines in nature, and it is not my intent to review all of them. Rather, this brief glimpse under the hood hopefully gives an impression of the key components required to make cells function. All cells have similar protein synthesis machinery. All cells have some basic energy-transduction machinery based on synthesizing ATP via a coupling factor. All cells possess some mechanism for donating and withdrawing electrons and protons to and from a hydrogen carrier. All cells create electrical fields across membranes that either produce or consume ATP. Finally, all cells on this planet are ultimately dependent on photosynthetic organisms, which convert solar energy to create the electrical fields that generate the flux of electrons and protons, making all life, including us, possible.

As we can see, the nanomachines that evolved in the earliest microbes allow cells across the tree of life to function. When viewing the legacy of ancient microbial nanomachines in the working of contemporary, living cells, one may get the impression that microbes have marched through the eons unchanged. This is not the case, though. As we return to microbes of the ancient world, we will see that they have evolved over time.

The first photosynthetic microbes were anoxygenic—that is, they were not capable of splitting water. It took several hundred million years before microbes evolved the ability to split water. Water is an ideal source of hydrogen on Earth's surface because it is far more abundant than any other potential electron donor, but splitting water takes a lot of energy. The responsible nanomachines evolved only once among prokaryotes: in the *cyanobacteria*, or blue-green algae. When these organisms finally were able to split water, they produced a new gaseous waste product: oxygen. The biological production of oxygen changed the evolution of life on Earth forever.

CHAPTER 5
||||||||

Supercharging the Engines

Oxygen is unique to Earth's atmosphere. The gas has not been found in high concentrations on any other planet in our solar system, nor has it yet been found in the surrounding neighborhood of stars that have planets. Although it is highly likely that other planets will be discovered to have oxygen, it does not seem to be a common gas on terrestrial planets.

The accumulation of oxygen was one of the most critical transitions in this planet's history—occurring long after life itself evolved—but the story about how Earth came to have oxygen in its atmosphere is complex. One chapter in that story is the evolution of the microbial nanomachines that produce oxygen. Although the evolution of the nanomachines was necessary for the production of oxygen, in and of itself it was not sufficient to allow the gas to become a major component of Earth's atmosphere. The oxygenation of Earth had much to do with chance and contingencies. As we will shortly see, oxygen became a major gas on Earth because of tectonics and the burial of the bodies of dead microbes in rocks. Once oxygen appeared in the atmosphere, it had a profound influence on the evolution of the microbes themselves and the cycles of elements that perpetuate life.

The history of how oxygen was discovered reveals an important property of the gas: it supports combustion. It had long been known that there was some component of air that allowed a flame to burn. In the eighteenth and nineteenth centuries, that property of air was used to detect oxygen. Originally, the gas was discovered by a German-Swedish pharmacist, Carl Scheele, in 1772. Retrospectively, Scheele's discovery was an amazing stroke of luck and insight. He had heated manganese oxide in a bell jar and observed that a product of the reaction made charcoal dust burn very fast. He repeated the experiment with mercuric oxide and got a similar result. He had no idea what manganese oxide or mercuric oxide were—to him they were just green and red minerals. But the invisible material that came from the minerals when they were heated and caused the charcoal to burn was very strange indeed. He called it "fire air" and wrote several letters about its strange properties. But Scheele wasn't a formal scholar and didn't write a scientific paper about his discovery until three years later. Consequently his experiments were not widely known.

In 1774, working independently, Joseph Priestley in England, conducted experiments similar to Scheele's using a magnifying glass to focus light from the Sun onto mercuric oxide. It is not clear whether Priestley knew about Scheele's experiments, but the results were similar. Rather than using charcoal, however, he placed a candle in the bell jar. The candle burned brighter and longer than one simply left in a bell jar with air alone. Moreover, with a flare of drama, Priestley showed that a mouse would live longer if exposed to the gas. (Obviously, readers should not try to replicate Priestley's experiments, because fumes of mercury are toxic.) Priestley also had no idea what the gas actually was, but he knew that plants could make the invisible material, which he called dephlogisticated air, on the basis of the now-archaic theory that substances that could burn contained an invisible substance, phlogiston. Priestley had placed a sprig of mint in a bell jar on a window sill and showed that after some time he could relight a candle with a magnifying glass to focus the sun's light in the closed jar. Without the sprig of mint, the candle would not light. But what was this invisible, odorless substance?

In late 1774, Priestley visited Antoine Lavoisier, a French nobleman, chemist, and tax collector who had a magnificent laboratory in Paris. Priestley described his experiments at a dinner party, where he probably had consumed a significant amount of wine. Lavoisier was intrigued and repeated Priestley's experiments to produce a "breathable" air by heating mercuric oxide. He appears to have been the third person to have made oxygen from a mineral, but he took another, more interesting and rigorous approach.

Lavoisier had a more sophisticated understanding of natural phenomena than Priestley and thought that if something was created by a chemical reaction, then something was also lost. The idea was simple but insightful—it formed the basis of what would be called quantitative analytical chemistry. It was not the beginning of chemistry, but rather, it was the beginning of an approach to the subject that allowed for rigorous testing of hypotheses. Being very rich, Lavoisier could pay the best French instrument makers to construct the finest equipment in the world at the time. Among the latter were extremely precise balances, built with meticulous attention to detail as if they were fine jewelry. They could measure changes in mass of 1 part in 400,000. Such precision was exceptional for the time, and Lavoisier used it to great advantage. By carefully weighing the mercuric oxide before and after heating, he could determine how much of the material was lost in the process. He then went on to do the reverse: he heated mercury metal in the presence of air, generating mercuric oxide, which weighed more than the original metal, and showed that the air in the chamber lost some volume. He repeated that experiment with phosphorus, producing phosphoric acid. Lavoisier also showed that the gas produced by heating mercuric oxide was a component of water and that the atmosphere of Earth was made primarily of nitrogen and this new component, which he called oxygen—"creator of acid." Lavoisier was the intellectual father of analytical chemistry and went on to discover several new elements before he was beheaded during the French Revolution, at the age of fifty, for his role in collecting taxes for the king.

Lavoisier did not understand how the oxygen could get into the atmosphere. It could have come from the Sun shining on rocks containing mercuric oxide or similar minerals, but that seemed improbable, because rocks do not appear to decompose when exposed to the Sun. Besides, when mercuric oxide is placed in a bell jar and simply exposed to light, nothing happens. One needed to heat the mineral to fairly high temperatures to get oxygen.

Part of the riddle was answered in 1779, when a Dutch physician, Jan Ingenhousz, working in the same laboratory in England in which Priestley had worked five years earlier, noted that aquatic plants produced bubbles on their green leaves when exposed to the Sun but not when kept in the dark. Sure enough, when painstakingly collected, the gases from the bubbles brought a smoldering candle to flame. Ingenhousz discovered that plants make oxygen, but neither he nor Lavoisier knew that it came from water.

As children, we all learn that plants make the oxygen we breathe, and most of us go on without thinking about the process further. But the fossil record indicates that terrestrial plants have been around on this planet for only about 450 million years. If the Earth is at least 4.55 billion years old, was there no oxygen before 450 million years ago?

As I described earlier, microbes had evolved a complex nanomachine capable of splitting water via the energy of the Sun billions of years before the rise of terrestrial plants, but it may be somewhat surprising that we still have a very unclear picture as to when the *first* microbe arose that was capable of doing so. There is only one extant prokaryotic group of photosynthetic microbes capable of producing oxygen: the cyanobacteria.

The evolution of cyanobacteria remains enigmatic. They are all closely related genetically and are the only prokaryotes that make the green pigment, chlorophyll *a*, which is used by all oxygen-producing organisms to split water. But perhaps most interestingly, they are the only photosynthetic prokaryote that has two different photosynthetic reaction centers. One is very closely related to the reaction center found in purple nonsulfur photosynthetic bacteria, but the latter are not capable of splitting water using energy from the Sun and consequently do not produce oxygen. They use light energy to split hydrogen gas into protons and electrons and subsequently make sugars. The other type of reaction center is derived from a photosynthetic green sulfur bacterium, the kind I was studying in the deep portion of the upper water column in the Black Sea. Those organisms also do not split water nor do they produce oxygen; they split hydrogen sulfide using light energy. Both purple nonsulfur photosynthetic bacteria and green sulfur bacteria are very sensitive to oxygen; if exposed to the gas they lose their photosynthetic capability. Somehow, the reaction centers from these two very different organisms appear to have found their way into one organism. How that happened is not clear, but it probably involved a series of gene-swapping events between different microbial species.

The resulting chimera, in which the two different reaction centers were genetically embedded within the nascent cyanobacterium, underwent further evolutionary modifications. A protein containing a quartet of manganese atoms was added to the reaction center from the purple bacteria; this would become the reaction center where water is split. Over time, the new cell modified the bacterial pigment system to produce chlorophyll, which allowed the reaction center to use light at higher energy levels for the water splitting. The second reaction center, a relic from the green sulfur bacteria, was also changed, and the modified nanomachine allowed it to work in the presence of oxygen. The resulting new mechanism, consisting of scavenged nanomachines, is extremely complicated, comprising over 100 proteins and other components in the two reaction centers, which work in series.

Let's return to our earlier metaphor about electrons as passengers in trains. In the first reaction center, light ultimately drives electrons from the hydrogen in water and moves them through a set of intermediate stations. The electrons arrive at the second reaction center, where, with the energy of light again, they are pushed very hard onto a packed train that then goes to another series of intermediate stations, where the electrons finally arrive at their destination. The destination is a small, ancient molecule called ferredoxin, which contains an iron and sulfur complex, similar to the mineral, pyrite, or fool's gold. There, through the aid of an enzyme, the electron finally meets its proton partner and forms NADPH. Recall that NADPH is a carrier of hydrogen and the hydrogen on the NADPH can be used to convert carbon dioxide to organic matter. The entire energy conversion machinery requires about 150 genes. It is the most complex energy-transduction machinery in nature.

It would appear that this machine, sometimes called the oxygenic photosynthetic apparatus, evolved only once in Earth's history. Because the production of oxygen changed the world so profoundly, my friend and colleague, Joe Kirschvink, at Caltech, whimsically called cyanobacteria "microbial Bolsheviks"—organisms that revolutionized the planet, but long before and more profoundly than the Russian Revolution.

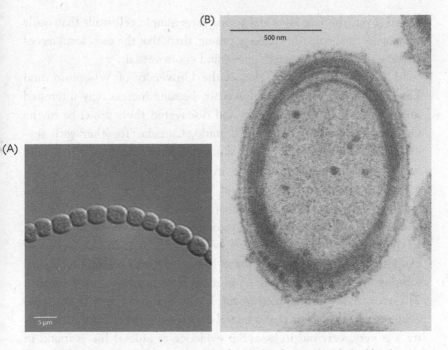

FIGURE 17. (A). A light microscope image of a chain of cyanobacteria (*Anabaena* sp.). (Courtesy of Arnaud Taton and James Golden) (B) A transmission electron microscope image through a section of a single cyanobacterium (*Prochlorococcus*) cell. The latter cell is approximately 1 micrometer in diameter and contains many membranes in which the photosynthetic apparatus (Fig. 16) and the coupling factor (Fig. 14 and 15) are embedded. In contrast with eukaryotic algae (Fig. 9), however, there are no membrane-enclosed organelles. (Courtesy of Luke Thompson, Nicki Watson, and Penny Chisholm)

These microbial Bolsheviks come in various shapes and sizes, from the very, very small *picoplankton*—cells that are just about 500 nanometers in diameter and that are so small as to be virtually invisible in a conventional light microscope—to relatively large cells that are linked together to form chains and are easily visible to the naked eye. In the contemporary ocean there are more than 1,000,000,000,000,000,000,000,000 (10^{24}) cells of cyanobacteria present at any given moment. But despite their numbers, it would be impossible to find such very small cells preserved in the fossil record. Even the largest cyanobacteria have simple cell walls that easily decompose. It should not be too surprising, then, that the early fossil record of these organisms is woefully sparse and controversial.

In the 1950s, Stanley Tyler, of the University of Wisconsin, and Elso Barghoorn, at Harvard University, became increasingly interested in microfossils in ancient rocks and discovered their presence in the Gunflint Formation in western Ontario, Canada. Together with several of his students—including William Schopf, Andrew Knoll, and Stanley Awramik—Barghoorn began to examine fossils from the oldest sequences in South Africa and Western Australia. As a student, Schopf was assigned by Barghoorn to work on samples from Western Australia and discovered a rich record that had never been reported. In the 1990s, Schopf, who by then was a professor at the University of California, Los Angeles, reported evidence of fossils resembling chain-forming cyanobacteria preserved in rocks from northwestern Australia. The rocks had been formed about 3.5 billion years ago. If true, the evidence would suggest that a microbe with oxygen-generating capability was very, very old indeed. But evidence of animal life is found in the fossil record only much, much later—about 580 million years ago. Could there actually have been almost a 3-billion-year lag between the evolution of microbes that produced oxygen, the cyanobacteria, and the rise of animals? If so, why?

Schopf's work was generally accepted, and he published several other papers with striking images of fossils that resemble the structures of cyanobacteria found in modern lakes. However, early in this century, a paleontologist at Oxford University in England, Martin Brasier, reexamined samples of the rocks Schopf had archived at the Natural History Museum in London and concluded that the fossils Schopf described were artifacts. Schopf's chains of cells, Brasier maintained, were not fossils of microbial cells at all but, rather, microscopic mineral deposits from underwater hot springs that formed structures that appeared to look like cells. The debate between these two camps has waged on; there is no agreement on the date of oldest physical fossils of cyanobacteria, although these organisms must have existed prior to the Great Oxidation Event (see below), approximately 2.4 billion years ago.

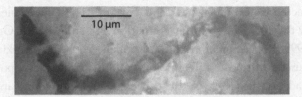

10 µm

FIGURE 18. Image of fossil microbes that resemble a chain of cyanobacteria (e.g., Fig. 17A) from the ~2.5 billion year old Gamohaan Formation of South Africa. (Courtesy of J. William Schopf, University of California Los Angeles)

Trying to circumvent the problem of preservation of the physical structure of microbes in rocks, chemists who work on the chemistry of rocks (geochemists) took another approach. In many cases, organisms die but relics of their bodies are preserved as chemical signatures in rocks. In fact, we intuitively know this, because petroleum and coal are the preserved remains of long-dead organisms. The proof that fossil fuels were formed from dead organisms came in 1936 when a German chemist, Alfred Treibs, showed that petroleum contained molecules that could only have originated from the plant pigment chlorophyll. Indeed, many of the people who work on the chemical signatures of organisms in the fossil record began their careers characterizing the organic components of petroleum for oil companies.

Although there are traces of other molecules preserved in sedimentary rocks, most of the chemical signatures are lipids—fats and oils—molecules that are not very soluble in water. For example, when animals, including us, die, one of the signature chemicals is cholesterol—a molecule that is in the membranes of animal cells but is not found in plants or prokaryotic microbes, such as cyanobacteria. But prokaryotic microbes make a set of molecules related to cholesterol, called *hopanoids*, which are part of their membranes. When they die, the hopanoids in the membranes of microbes are sometimes preserved in the rocks for billions of years. In fact, it is sometimes said that hopanoids are the most abundant naturally occurring organic molecules on Earth.

Cyanobacteria make a relatively specific hopanoid, and the degradation products of the molecules can be preserved in rocks if they are not subjected to too much heat and pressure. These molecules are not found in the rocks in the Isua Formation in southwestern Greenland, but in 1999, Roger Summons, an Australian geochemist working at the Massachusetts Institute of Technology, and his colleagues reported that the degradation products of the particular molecule found in modern cyanobacteria was present in the Pilbara Craton in Western Australia (close to the same place Schopf had worked). Those rocks are dated to 2.7 billion years before the present. While there is still controversy surrounding the origins of cyanobacteria, the molecular data seemed to suggest that these organisms evolved at least 2.7 billion years ago, possibly earlier. But the lipid analysis has also been questioned. Some of the biomarkers may be contaminants from the oils used during the drilling process to collect the samples. Indeed, the field seems to be going in circles. Increasingly, the evidence of microfossils from about 3.5 billion years ago is accepted with caution; whether these were cyanobacteria or not is unclear. But what is clear is that for approximately the first 4 billion years of Earth's history, there are no signs of animal life. If animals require oxygen, and oxygen required the evolution of cyanobacteria, then when did cyanobacteria finally produce enough oxygen to have an impact on Earth's atmosphere? It is now known with reasonable certainty that it was between about 2.3 and 2.4 billion years ago. However, the evidence for this date is a bit arcane.

There are four stable isotopes of sulfur in nature, and it is their distribution in rocks over the past 3.5 billion years that provides a basis for understanding when Earth's atmosphere became oxygenated. The lighter isotopes of sulfur, which contain fewer neutrons, vibrate faster than the heavier isotopes. Because of the increased vibrational frequency of the lighter isotopes, they tend to collide more often with neighboring atoms, and hence they have a higher chance of forming chemical bonds with other elements than the heavier isotopes do. Using a mass spectrometer, a machine that can very precisely measure the abundance of the different isotopes, in 2000 James Farquhar, Huiming Bo, and Mark Thiemens showed that the isotopes of sulfur in sedimentary rocks have a very unusual pattern. In rocks older than about 2.4 billion years, including those from Australia that contain the hopanoid biomarkers of cyanobacteria, the isotopic composition of sulfur is rather haphazard; there is no pattern of the isotopes based on their masses. However, from 2.4 billion years to the present, the isotopic composition is clearly based on the number of neutrons in the element. That is, they behave as predicted by their mass: the heavier isotopes of sulfur, with more neutrons, are less abundant in the minerals in the rocks than the lighter isotopes. Something happened about 2.4 billion years ago that changed the way sulfur isotopes formed bonds. But how does that tell us anything about oxygen?

Much of the sulfur that is found in rocks originally came from volcanoes in the form of a gas, sulfur dioxide (SO_2). Sulfur dioxide is a colorless gas with a sharp odor; you can smell it for miles around paper factories because sulfur-containing molecules are often used to break down wood to make pulp for paper. The bonds in sulfur dioxide can be broken by the high-energy, ultraviolet light emitted from the Sun. When the ultraviolet light breaks bonds, it does not discriminate between isotopes. The resulting products have the same isotopic composition as the starting materials.

Ultraviolet radiation is not visible to human eyes but causes our skin to burn and can cause mutations in our skin cells that can lead to cancer when we are exposed to too much of it. Although some ultraviolet radiation from the Sun hits the Earth's surface in the modern world, most of it does not. It is absorbed high up in the atmosphere, in the stratosphere, by another gas, which is composed of three atoms of oxygen. That gas is ozone. The only known mechanism to produce stratospheric ozone on a planet requires some free oxygen in the atmosphere.

The change in the pattern of the distribution of sulfur isotopes in rocks is interpreted as the development of a layer of stratospheric ozone about 2.4 billion years ago. That explanation requires that oxygenic photosynthesis by cyanobacteria ultimately led to an increase of oxygen in that atmosphere. The sulfur isotopic record clearly suggests the world went through a key transition: before 2.4 billion years ago, there was virtually no free oxygen in the atmosphere; after 2.4 billion years ago, there was. Geologists poetically (and somewhat dramatically) named the transition the Great Oxidation Event. This "event" took place during a 100-million-year, or longer, period. It appears to have been a singularity in Earth's history—that is, it happened only once. We can conclude that because the sulfur isotopes in the rock record from 2.4 billion years ago to the present are faithfully fractionated according to the masses of their isotopes, but prior to 2.4 billion years ago, the fractionation of sulfur isotopes was independent of their masses. This interpretation of the sulfur isotopes suggests that oxygen has been part of this planet's atmosphere for the past 2.4 billion years. The concentration of oxygen right after the transition was quite low—probably less than 1% of the current level—and wasn't sufficient for the evolution of animals.

It takes more than the evolution of a photosynthetic nanomachine to endow a planet with oxygen in its atmosphere. In order for the gas to become abundant, huge amounts of the microbes that had the photosynthetic nanomachines had to die and subsequently become incorporated into rocks. The deaths of the photosynthetic microbes over hundreds of millions of years ultimately paved the way for our very existence. Let's take a look at that apparent paradox—why the death of the cells that make oxygen is necessary for oxygen to become an abundant gas.

Consider the oxygen we are breathing now. The concentration of oxygen in Earth's atmosphere has been unchanged in our lifetime and the lifetime of our great, great, great, great, great (and you can add a lot more "greats") grandparents. It comprises 21% of the volume of air on Earth and has been extremely constant over hundreds of thousands, if not millions, of years. How do we know? Because we can measure the oxygen trapped as bubbles of gas in ice cores drilled from the Antarctic ice sheets and determine with great precision and confidence that the oxygen concentrations have remained fundamentally unaltered over the past 800,000 years. During that time, the production of oxygen by all the algae and plants on Earth was balanced by the consumption of oxygen by respiration of all the animals and microbes. For the concentration of oxygen to change in Earth's atmosphere, something had to disrupt the balance between photosynthesis and respiration.

There were no plants or animals 2.4 billion years ago. Indeed, there were only microbes. All life on Earth was basically confined to the oceans and other watery places. The photosynthetic cyanobacteria, with their oxygen-generating nanomachinery, didn't make oxygen for the sake of making oxygen. Oxygen is a waste product of the photosynthetic process.

Organisms split water to get the hydrogen, and they use the hydrogen to produce organic matter. Oxygen is oxidized water, and effectively, organic matter is reduced carbon dioxide and nitrogen gas. Organic matter contains energy but also can be used to make sugars, amino acids, lipids, and nucleic acids; in short, organisms use the organic matter to make another cell. For want of a simpler term, I'll call the organic matter that the cells make "cell stuff." In effect, photosynthesis transfers hydrogen from water to carbon dioxide and nitrogen to form cell stuff, which the cells accumulate and which eventually allows them to replicate. In respiration, organisms use the organic matter to make energy without sunlight and to make other cells. Respiration strips the hydrogen from the carbon and adds it to oxygen, releasing water and carbon dioxide as waste products. We intuitively know this when we breathe on a cold window—water vapor condenses. Our respiration has added hydrogen from the food we eat to the oxygen in the air we breathe to form water. In essence, the planet runs on a cycle of water-splitting by photosynthesis to form oxygen and the production of water by respiration.

To get oxygen to accumulate in the atmosphere in large quantities, some of the cell stuff produced by the photosynthetic microbes has to be hidden from the respiring microbes. It is sort of like trying to hide candy from children. If you want to hoard candy, then you have to find a place where the children can't find it. One place to hide the candy is to put it way out of reach on the top shelf in the back of a dark closet. The back of the dark closet of Earth is in rocks. Microbes have a hard time respiring the organic matter in rocks—not that they don't try.

A very small number of phytoplankton, including cyanobacteria, sink to the bottom of the sea. The actual fraction that makes it to the sea-floor varies with depth of the ocean: the deeper the water column, the smaller the fraction that reaches the bottom. In the contemporary oceans, virtually no organic carbon reaches the bottom in water columns deeper than about 1000 meters, which means that the modern deep oceans do not store organic carbon. By far the most important storage areas are in shallow seas and along the coasts of continents. But even there, on average less than 1% of the organic matter produced by phytoplankton reaches the seafloor, and only about 1% of that is subsequently buried in sediments. That means that less than 0.01% of the organic matter is actually buried, but over millions and millions of years, this very small fraction becomes significant on a global scale. The cell stuff from the dead organisms becomes mixed into the sediments, and as younger sediments pile up over older sediments, the decomposing body parts of the dead microbes get squeezed and heated and ultimately become incorporated into sedimentary rocks—rocks derived from the erosion of other rocks on land. Some of the sedimentary rocks, which contain the organic matter, are subsequently uplifted onto the continents to form mountains. Without burial, the organic matter would be respired and little or no oxygen would accumulate. Without uplifting the organic matter onto continents, the organic matter would be subducted into the interior of the Earth by tectonic processes, heated, and returned to the atmosphere as carbon dioxide from volcanoes—and little or no oxygen would accumulate. So, as the organic matter in the sedimentary rocks that were preserved on continents slowly accumulated, the oxygen concentration in the atmosphere slowly rose. It took a long time, but without the entire process, we would not be breathing oxygen.

One of the curious issues about the Great Oxidation Event is why it took so long for it to happen—or did it? If the incredibly sophisticated nanomachines capable of splitting water evolved in cyanobacteria evolved just prior to 2.4 billion years ago, then they transformed the planet within 100 million years or less. However, if they evolved much earlier, as suggested by the fossil record, why did it take hundreds of millions of years more for oxygen to become a significant gas in Earth's atmosphere? The answer is not easy to understand, and all explanations to date are controversial.

For a long time I thought that the lag of hundreds of millions of years between the evolution of cyanobacteria and the rise of oxygen in the atmosphere was due to the reactions of oxygen with iron and sulfides in the Archaean ocean more than 2.5 billion years ago. Oxygen is the most abundant element in the Earth's crust, but not as a free gas. Oxygen is very promiscuous and doesn't like to be alone. It is a very reactive molecule and chemically combines with many metals and other elements. If you put a nail in well-aerated water for a few days, it will form rust,

FIGURE 19. A photograph of a section of black shale formed about 185 million years ago. This period of time (the lower Jurassic) witnessed very high productivity in the oceans and the subsequent burial of carbon in sediments. (Courtesy of Bas van de Schootbrugge)

which is iron combined with oxygen—iron oxides. Three billion years ago, the oceans contained large quantities of dissolved iron, and during the next several hundred million years after the evolution of the oxygen-splitting nanomachines, iron oxides (rust) precipitated in many regions of the ocean. The reaction of oxygen with iron proceeded for almost two billion years and did not require any biological intervention. Iron will rust with or without the presence of microbes; the only ingredients required are oxygen and water. While the oxidation of iron consumed oxygen, some back-of-the-envelope calculations suggest that this process alone could not have hindered the rise of the gas in the atmosphere for hundreds of millions of years. Something else had to impede the accumulation of the gas.

The production of oxygen created opportunities for the evolution of new metabolic pathways for microbes. These new opportunities led to changes in the distribution and abundance of several other elements, especially sulfur and nitrogen. Before the production of oxygen, most of the sulfur in the oceans was in the form of that gas with the rotten egg smell, hydrogen sulfide, which was, and continues to be, supplied to the ocean from deep-sea volcanoes, called *hydrothermal vents*. The water flowing from these vents is extremely hot, about 300° Celsius, and contains large amounts of sulfides and iron, which when cooled, form mineral chimneys of "fool's gold," or pyrite. In the presence of oxygen, some microbes evolved a new set of nanomachines that allowed them to take the hydrogen from the hydrogen sulfide and use it to fix carbon dioxide to make organic molecules. The oxygen provides an electrical gradient between the electron-rich fluids and gases coming out of the vents, and the electron-poor gas, oxygen, and other molecules in the ocean waters surrounding the vents. This electrical gradient supplied the driving force for a new kind of metabolism. Unlike the photosynthetic green sulfur bacteria, such as those in the Black Sea, these sulfide-oxidizing microbes in hydrothermal vents can split hydrogen sulfide without using energy from the Sun directly. Their carbon-fixing machinery is virtually identical to that found in cyanobacteria, but the metabolic innovation, which is called *chemoautotrophy* (chemical self-feeders), allows carbon fixation to occur in the deep dark interior of the ocean, but only because oxygen is produced by cyanobacteria in the sunlit portion of the ocean hundreds and thousands of meters above.

The basic concept is that if hydrogen is directly bound to oxygen, such as occurs in water, it takes a lot of energy to remove the hydrogen. The only source of energy that is used to remove the hydrogen biologically is in the visible light spectrum of the Sun. The hydrogen bound to sulfur is far easier to remove than from water. It takes only about 10%

of the energy to pull the hydrogen from sulfide than water, but in the presence of oxygen, the sulfur can be converted by microbes to form an oxide—sulfate, an atom of sulfur containing four oxygen atoms.

Is the microbial oxidation of hydrogen sulfide the key reaction that could explain the lag in the rise of oxygen? For a long time, I thought that was possible. But when we began to learn a bit more about the supply of sulfur to the ancient oceans and made some simple calculations, it seemed increasingly improbable that it was the major cause of the delay in the rise of oxygen. The rise of oxygen could have turned all the iron into rust and oxidized all the sulfides to sulfates, but it should never have taken 300 million years or more to get oxygen to be abundant in the atmosphere. Something was wrong. Once again, experiments in the Black Sea provided a clue.

In the Black Sea, there is a place in the water column where oxygen becomes nonexistent and hydrogen sulfide becomes increasingly abundant. It took me a few years to understand how this transition in the chemistry of the Black Sea reflected the chemistry of the Earth and the rise of oxygen. Even if the deep water of the Black Sea is only 1500 years old, there was a transition in microbial metabolism from the upper, oxygenated water to the deep water. It was as if I were going back in time to the Great Oxidation Event.

The most abundant gas on Earth is nitrogen, but it is in a form that is extremely chemically stable. The nitrogen gas in our planet's atmosphere is composed of two atoms of the element, bound to each other by three chemical bonds. In contrast to oxygen, nitrogen gas (N_2) is virtually inert. If the Earth had only nitrogen in its atmosphere, the newspaper on your sidewalk would never turn yellow and decompose, iron would never turn to rust, and a candle could not burn. But unless some hydrogen is added to the nitrogen, life on Earth would not exist because no microbe can make amino acids or nucleic acids without nitrogen bound to hydrogen. Fortunately, microbes can attach hydrogen to nitrogen, but it takes a lot of energy.

I realized that the cycle of nitrogen, which is totally dependent on microbial activity, behaves in almost exactly the same way as the cycle of sulfur. Nitrogen is required to make proteins and other critical molecules required by cells. But to get nitrogen into cells, organisms have to either acquire it from their environment as an ion or somehow chemically alter the gas in the atmosphere. Long before the existence of oxygen as a gas on Earth, some microbes evolved that could add hydrogen to the nitrogen in the atmosphere (or dissolved in water) with aid of a complicated, extremely ancient nanomachine, an enzyme called nitrogenase. The product of that reaction is ammonium. Ammonium is a single atom of nitrogen with four hydrogen atoms bound to it (NH_4^+). In the absence of oxygen, ammonium is extremely stable, but when oxygen became available, microbes evolved another set of machines that allowed them to strip the hydrogen from ammonium and use the hydrogen to convert carbon dioxide into organic matter without using energy from the Sun. Like their deep-sea counterparts, these microbes are also chemoautotrophs: they grow by using the electrical gradient between an electron-rich molecule, ammonium, and an electron-poor molecule, oxygen. These ammonium-oxidizing microbes can't make a living unless there is free oxygen present in their environment. The products of their reactions are forms of nitrogen that contain oxygen, particularly nitrate (NO_3^-), which is a nitrogen atom with three atoms of oxygen bound directly to it. Just as in the case of sulfur, when no oxygen is present other microbes can use nitrate in respiration; however, unlike the case with sulfur, anaerobic respiration with nitrate doesn't lead to the formation of a molecule with hydrogen attached to it, such as ammonium. Rather, it leads to the production of nitrogen gas.

Looking at the chemical compounds of nitrogen in the Black Sea, it is clear that in the upper, oxygenated area of the water, nitrate is abundant and there is no ammonium. However, deeper in the water column, where there is no oxygen, and hydrogen sulfide is abundant, ammonium becomes the only form of fixed nitrogen. But a careful consideration of the vertical distribution of oxygen and hydrogen sulfide in the Black Sea gave me pause. At the point where oxygen becomes increasingly scarce and yet hydrogen sulfide is hardly present, there is virtually no nitrate or ammonium. That is a place where microbes find it very hard to live. The cyanobacteria that generated the oxygen in the early oceans would have helped make it possible for other microbes to use the oxides of nitrogen in respiration, but unlike the sulfur cycle, the product of the respiratory

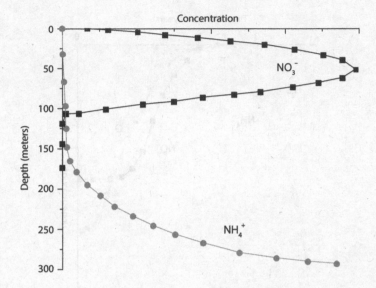

FIGURE 20. A vertical profile of the distribution of two forms of nitrogen, nitrate (NO_3), and ammonium (NH_4), in the Black Sea. Note that where oxygen becomes vanishingly low (Fig. 1), those forms of nitrogen also become extremely scarce.

reactions with nitrogen wasn't an ion, like sulfate—it was two gases, and went back into the atmosphere. The nitrogen cycle, driven totally by microbes, prevented the planet from having oxygen for a long time. Indeed, work with my colleagues at Rutgers, especially Linda Godfrey, suggests that for at least 300 million years before the Great Oxidation Event, the cyanobacteria had produced oxygen, which was ultimately used by other microbes to convert ammonium to nitrate, and then on to nitrogen gas, with the result that the ocean lost fixed nitrogen. Without the fixed nitrogen, phytoplankton would not be able to make a lot of cell stuff and organic carbon could not easily be formed. If organic carbon can't be easily formed, it can't be buried. If organic carbon can't be buried, oxygen could not accumulate in the atmosphere. In effect, it appears as if the entire microbial system of the early oceans was rigged by natural feedbacks to stay anoxic. Life almost certainly evolved under anoxic conditions, and the metabolism of microbes appears to have kept the planet anoxic for the first half of Earth's history. At some point,

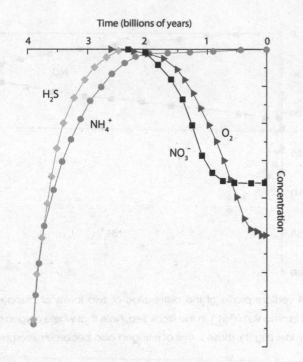

Time (billions of years)

FIGURE 21. By turning the vertical profiles of oxygen, nitrogen, and hydrogen sulfide on its side, one can imagine the progression of how the chemistry of the ocean changed prior to the Great Oxidation Event, ~2.4 billion years ago, and after the oxygenation of the atmosphere and ocean.

N_2 and N_2O (nitrous oxide, or laughing gas) were produced. Both gases escaped the oceans however, around 2.4 billion years ago, cyanobacterial production of oxygen finally overtook the consumption of the gas by other microbes, and the atmosphere finally became oxidized. Perhaps amazingly, we really don't know for certain how it happened.

The evolution of this planet, containing oxygen in its atmosphere, was the culmination of hundreds of millions of years of evolutionary innovation of nanomachines that ultimately harnessed solar energy to split water. But the rise of oxygen also had a profound impact on the evolution of many microbes themselves.

Being a highly reactive gas, oxygen can be both a wonderful, but also dangerous, place to put hydrogen in respiration. It is wonderful because the reaction of hydrogen and oxygen allows a lot of energy to be derived. Indeed, if you light a match to a mixture of hydrogen and oxygen gases, you will create a violent explosion. These two gases are literally rocket fuels. The world of oxygen is a high-energy world. Microbes that could use oxygen for their respiration had to make only relatively small changes to their respiratory machinery to ensure that when they combined the hydrogen from their respiration of cell stuff it didn't react so violently with oxygen that the cells literally burned up. The control of that reaction required the evolution of another nanomachine, one that very deliberately couples the electrons and protons to oxygen. The energy of that reaction is enormous, and from it microbes can generate eighteen times more ATP for each sugar molecule they respired than with the ancient anaerobic respiratory system. We have adopted exactly that process in our own subcellular power-generating nanomachines—the mitochondria. The production of oxygen literally led to the supercharging the engines of life!

The evolution of nanomachines was also critical to the development of cycles of elements that perpetuate life on Earth to this day. Because of the heat generated by the radioactive decay of elements deep in the Earth's interior, the elements essential for life are continuously replenished from gases emitted by volcanoes, by the weathering of rocks, and from the burial of the body parts of microbes. This process has been ongoing since the formation of the planet, 4.55 billion years ago, and will continue for several billion years into the future. However, the evolution of microbial nanomachines and the subsequent rise of oxygen changed how these elements were cycled on a planetary scale. Specifically, the evolution of microbial nanomachines allowed organisms across the planet to become connected via their internal machinery in a giant electron circuit. This circuit is based largely on transfers of hydrogen to and from four of the six principal elements: carbon, nitrogen, oxygen, and sulfur.

To connect the metabolism between organisms requires some kind of "wires," and the ocean and the atmosphere are the two major "wires" on Earth. We don't even have to leave our seats to see how that works.

Take a deep breath. The oxygen you just inhaled wasn't produced in the room you are in now. There is no giant standing outside focusing the energy of the Sun on metal oxides with a huge magnifying glass, nor are we carrying around a culture of algae on our backs. We breathe oxygen in the winter, even if there are no plants photosynthesizing in our immediate neighborhood. The oxygen we breathe was produced perhaps a million years ago and is brought to you from far away, courtesy of the atmosphere. A long time ago, some plants and phytoplankton, somewhere on Earth, produced the oxygen that you and I are breathing. We are living off the kindness of strangers. But our respiration produces carbon dioxide and water—a very weak seltzer (which also was invented by Priestley). The carbon dioxide we exhale is used by phytoplankton and plants to make more plants and phytoplankton elsewhere on the planet.

The ocean also connects Earth's metabolism. The currents in the ocean bring the oxides of nitrogen to the surface, where phytoplankton consumes them to make new cells, some of which sink deep into the ocean and provide a source of food and energy for microbes and other forms of life in the ocean interior. Because the ocean is a huge interconnected fluid that circulates on a global scale, the water deep in the ocean interior ocean gets oxygen from the atmosphere. In two major regions of the ocean, the North Atlantic off Greenland and in the Antarctic, very cold waters are formed during the respective winters. Cold waters are dense and tend to sink; water has a maximum density at 4°C. The colder the water is, the more it can absorb oxygen. The cold, dense, oxygen-rich waters carry the gas all across the ocean in a slow *conveyor belt* from the Atlantic to the Indian Ocean across to the Pacific and back. The round trip takes about a thousand years. That conveyor belt allows the microbes in the ocean interior to use the sulfide or ammonium to fix carbon based on the oxygen produced millions of years ago, somewhere far, far away. When oxygen finally became available and coupled the biological cycles of sulfur, nitrogen, and carbon, it potentially was also responsible for a major change in Earth's climate and possibly the first mass extinction on the planet.

There is compelling evidence that about 200 million years after the Great Oxidation Event, massive ice sheets formed in several parts of the world and lasted about 300 million years. It was the longest and possibly one of the most extensive glaciations in Earth's history. Ice was present not only on land but also throughout the oceans, possibly even covering the oceans to the equator, a so-called snowball Earth. What triggered such a massive change in climate?

One of the possible causes of this climate shift was the accumulation of atmospheric oxygen. While the interior of Earth is heated by radio-activity, the surface of this planet is heated by the Sun. Solar radiation ultimately is reflected back to space, but some of it is trapped by the blanket of gases in Earth's atmosphere. At the present time, the most important gases that trap the heat are water vapor and carbon dioxide. In fact, without these greenhouse gases in the atmosphere, Earth's oceans would be frozen. But the situation was even more extreme 2.4 billion years ago. At that time, the Sun was approximately 25% less bright than it is at present, which means it gave out less heat. In order for the oceans to be liquid at the surface, greenhouse gases had to be very abundant and very good at absorbing solar energy, especially infrared radiation—a type of energy we cannot see with our eyes but can feel with our skin; infrared radiation is heat. One of the most efficient gases that absorbs infrared radiation is methane.

At present, methane is a relatively minor greenhouse gas, but 2.4 bil-lion years ago, it almost certainly was far more abundant. Methane is a very simple gas: it is composed of one carbon atom bound to four hydrogen atoms (CH_4). It burns extremely efficiently in the presence of oxygen, which means that the gas has a lot of energy stored in its bonds. Methane is made as a product of respiration by some microbes under strictly anaerobic conditions. That is, if oxygen is not available, some microbes can use a special nanomachine to put the hydrogen from sugars and other organic molecules onto carbon dioxide to form methane. These microbes are the Archaea—the second large group of prokaryotes that Woese and Fox had discovered. The nanomachinery in the methane-producing microbes is very sensitive to oxygen; small concentrations of oxygen effectively stop the microbial production of methane. Methanogenic microbes are commonly found in many places today, including the guts of cows and other ruminants, as well as in about

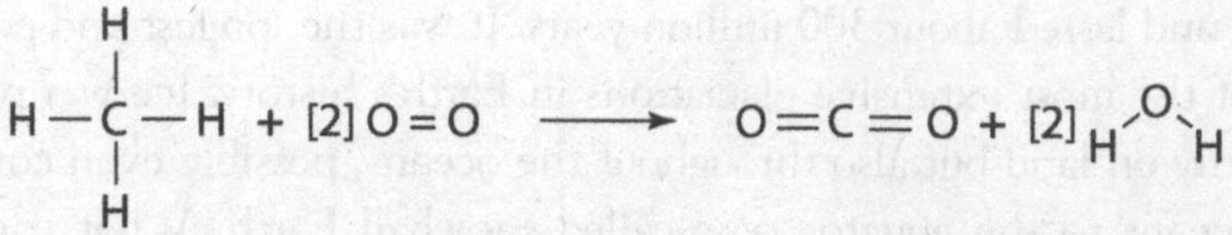

$$H-\overset{\overset{\displaystyle H}{|}}{\underset{\underset{\displaystyle H}{|}}{C}}-H \; + \; [2]\,O=O \;\longrightarrow\; O=C=O \; + \; [2]\,\overset{\displaystyle O}{\underset{H}{}}\overset{}{H}$$

Figure 22. A schematic showing the difference between methane (CH_4) and carbon dioxide (CO_2). Both of these molecules are invisible, odorless gases. In the presence of oxygen, methane is converted to CO_2 and water in the atmosphere and by microbes.

40% of humans. But 2.4 billion years ago, these organisms would have been extremely abundant in the coastal waters of the world.

Even in the presence of oxygen, several other kinds of microbes can use methane as a source of energy and for cell growth. The consumption of methane by microbes is one of the fastest and most efficient ways of destroying the gas. Once that capability evolved, the methane-destroying machinery must have dramatically reduced the flux of methane from the oceans to the atmosphere, and the oxygen in the atmosphere would have, with the aid of sunlight, destroyed atmospheric methane. A major infrared-heat-absorbing gas was gone, and the faint younger Sun could not provide enough heat to keep the oceans from freezing. The ensuing formation of ice or icy slush across the ocean surface almost surely reduced the area for the growth of photosynthetic microbes and simultaneously would have impeded the exchange of gases between the ocean and atmosphere. The geological record suggests that several extensive periods of cold, inhospitable oceans ensued. Kirschvink, who called the cyanobacteria microbial Bolsheviks, further whimsically dubbed the condition of global ice sheets across the oceans "snowball Earth." If this scenario is true, it was the first time in the geological history of Earth that microbes completely disrupted the climate of the planet.

Snowball Earth conditions seem to have happened several times, the last being about 750 million years ago. Incredibly, in all cases, the instructions for making all the basic nanomachines appear to have been passed on to the small number of surviving microbes. These organisms were pioneers, ferrying life across vast swaths of planetary destruction.

CHAPTER 6

IIIIIIII

Protecting the Core Genes

Life on Earth is precarious, inevitably transient, yet incredibly durable. From time to time, catastrophic events far beyond the control of any living organism lead to massive losses of species. The fossil record during the past 550 million years reveals at least five major extinctions of marine animals. With one exception, the causes are poorly understood. That exception occurred 65 million years ago and almost certainly resulted from the impact of a large meteor that struck near the present-day area off the Yucatan coast in Mexico. It was a bad day for dinosaurs and many plants. But microbes sailed right on through that extinction, just as they had in all the other previous extinctions that span vast times in Earth's history. How did nature ensure that the instructions for manufacturing the core nanomachines would persevere through the extreme traumas that would kill massive numbers of animals and plants?

The instructions for replicating the core nanomachines are encoded in genes. Genes are sets of sequences composed of four molecules, deoxyribonucleic acids, used by all organisms as instructions for making proteins. In prokaryotes, such as bacteria, several million deoxyribonucleic acids are strung together to form a large circular molecule that contains the instructions for making several thousand proteins. The proteins are, in turn, composed of twenty individual amino acids strung together in a specific order. The twenty amino acids to make proteins are found in every living organism on Earth.

Sets of three deoxyribonucleic acids in a specific order encode a specific amino acid, and the proteins are manufactured on those very old nanomachines, the ribosomes. The proteins are themselves used to make the nanomachines that allow the organism to generate energy and reproduce. The reproduction of cells is dependent on replication of the genes, and the replication of genes is dependent on the ability of the organism to generate energy, survive, and grow.

The basic discovery that genetic information is inherited is attributed to Gregor Mendel, an Austrian monk who examined the pattern of colors of flowers and seeds, shapes of pods, and so on in some 29,000 samples of peas. His work was published in German in 1865, six years after Darwin published the first edition of *The Origin of Species*. Obviously Darwin could not have known about genes. In fact, Mendel's work was largely ignored until the early part of the twentieth century, when it was rediscovered and given a figurative life by a British biologist, William Bateson, who coined the term *genetics*. Bateson himself had no idea how genetic information was transmitted from generation to generation, but he recognized, based on Mendel's work, that there were basic, predictable patterns in the offspring of mating pairs. It wasn't until the second half of twentieth century that it was realized that nucleic acids were responsible for carrying the instructions for how proteins are made and how the patterns are dictated.

One of Darwin's major aha! moments was when he realized there was natural variability within a species that could be selected by breeding. For example, humans had clearly used the natural variations in dogs to breed new forms with new traits; but they were all dogs. If humans could do this with dogs, or horses, or pigeons, why couldn't nature? At that time, there was a clear definition of a species: a species, in the context of animals and plants (and that was all that mattered at the time), was an organism that could sexually reproduce viable offspring—that is, the offspring also could sexually reproduce. Pigeons can reproduce viable offspring when mated with other pigeons, but the offspring of a cross between a pigeon and an eagle, if it were viable, could not reproduce. A cross between a male donkey and a female horse yields a mule, which is a sterile animal. Pigeons and eagles, horses and donkeys are all different, identifiable species.

The variations in species, Darwin asserted, are selected by competition within the species, gradually leading to changes such that a new species can no longer reproduce with the last ancestor of the new species and form viable offspring. This notion—of descent with variation, followed by selection and speciation—forms the theoretical basis of Darwinian evolution. That genes are transferred from parent to progeny, or by descent, is a concept of *vertical* inheritance. In organisms that primarily replicate via sexual recombination, that is the way genes are passed on. But this isn't the only mode of transmission of genes across large swaths of time, especially in microbes. Before going into microbial evolution and how the nanomachinery is passed on, let's turn to the issue of why there is variation within species to begin with, because without variation, there could not be evolution as we know it.

From time to time in the process of replication of genes, a cell makes a mistake, and the copy of the gene is slightly altered from the original. Much like monks copying books, mistakes are almost always a mismatch in the "spelling" of the sequences of nucleic acids during replication. There are four nucleic acids in DNA: adenine, guanine, cytosine, and thymine—which are abbreviated A, G, C, and T. DNA is composed of two strands, and for each T on one strand there is an A on the other strand. Similarly, each C on one strand is matched with a G of the other. However, in the presence of high levels of ultraviolet radiation, for example, there is a higher-than-average probability that rather than matching a T with an A, the energy from the radiation causes a T to be matched with a nearby T on the opposite strand. Unless that mutation is repaired, the organism will carry it onward.

Many other types of single nucleotide mutations can occur, and most of these mistakes don't fundamentally alter the ability of the cell to grow and replicate. As we discussed earlier, these types of mistakes (neutral mutations) can lead to variations, but do not confer any advantage or disadvantage on the organism. Some people have blue eyes and some have brown eyes; some have curly hair and some have straight hair; some have large noses, some small. These variations have little or no influence on the ability of a human to reproduce and simply are present because of the minor genetic "mistakes," or variations, that are carried on in a population. By definition, neutral mutations do not influence the ability of the organism to replicate and produce viable offspring; the mutations simply are passed on from generation to generation.

However, some mistakes can be very detrimental. In humans there are many mutations of a single nucleotide that can lead to very serious, and sometimes lethal, diseases, such as cystic fibrosis, hemophilia, and Tay-Sachs disease. In such cases, the carriers seldom live to a reproductive age, or if they do, they often cannot reproduce. Similarly, in microbes, these single-nucleotide, or *point*, mutations that lead to, for example, the inability of a cell to make proteins, respire, or make ATP efficiently, inevitably will lead to the death or extinction of that organism. They are not carried on.

There are several other kinds of mistakes besides point mutations of nucleotides. Organisms sometimes mistakenly make serial duplicates of genes, called *tandem repeats*, which gives rise to a pair of proteins that is duplicated and stuck together. This gene-duplication process is sort of a molecular version of Siamese twins that cannot be separated. In other cases, bits and pieces of genes are mistakenly inserted in the middle or at the ends of another gene. The outcome may lead to variations in the lengths of the protein, but if the core machinery works, the genes for that new protein may be retained. In many cases, this type of mutation can lead to new functions for the genes.

Mistakes continuously and spontaneously occur in all genes in all organisms, and sometimes they are beneficial. If the mistake allows an organism to outcompete others in acquiring energy or expanding the range in which it lives while still reproducing viable offspring, it is said to confer a *selective advantage*. Many genes, it would appear, explore the limits of diversity through mutations. That is, many of these very divergent genes are successfully passed on from generation to generation and are retained so long as they are advantageous, or at least not disadvantageous, to the organism.

The outcome of these continual random mistakes has led to huge numbers of variations in genes, and almost all of the variations are in microbes. It is estimated that at any moment in time, there are approximately 1,000,000,000,000,000,000,000,000 (i.e., 10^{24}) microbes living on Earth. That is an enormous number of self-replicating organisms.

To put it in some perspective, the number of extant (living) microbes is about 100,000 times more than all the stars in the visible universe. Each microbe contains approximately 10,000 genes. Humans have, through the technologies of gene sequencing and computational searches, identified over 25,000,000 genes in nature, with millions more added each year. We don't really know how many genes there are on Earth, and the number may not be knowable, because genes are continuously changing. Assessing the population of genes is something akin to trying to count the number of raindrops that fall on the planet each day. A reasonable estimate is probably on order of about 60 million to 100 million genes.

Approximately 40% of the genes that have been identified have no known function. That they are retained in an organism almost certainly means that these genes work in some way—we just don't know how. The other 60% have presumed functions based on their similarity to genes that have been identified as working in a specific way in some organism. In a classical case of Darwinian selection, each gene would randomly mutate over time, optimizing its function to facilitate the organism carrying the gene to more efficiently acquire resources and replicate. But it doesn't quite work that way.

Not all genes are created equal. Although most genes do mutate, changing slowly over time and varying from organism to organism, the genes that encode for very highly specialized components of key nanomachines hardly change at all. For example, in photosynthetic organisms, the various proteins that form the core structure of the machine have to fit and work together, and they also have to hold other components in specific positions and orientations, otherwise the machinery would not function. Each of the proteins that compose the core structure of the machine is encoded by a specific gene. Careful inspection of these genes reveals that they are virtually identical, from the oldest extant oxygen-evolving photosynthetic organisms, the cyanobacteria, to the most recently derived land plants. In fact, one of the major proteins, called D1, which is found in the photosynthetic reaction center that splits water, is 86% identical across all oxygen-producing photosynthetic organisms. This doesn't mean that there are no mistakes made when the genes for D1 are copied, but it does mean that very small mistakes often result in fatal outcomes for the organism that inherits the mutant genes. The lack of variation in the genes coding for these nanomachines indicates there is an absolute necessity for the proteins to be precision made so that all the parts fit in an extremely accurate way; otherwise, the machines do not function properly.

Many of the proteins that form structural components in the core machines have similarly small variations. These include the machines responsible for respiration, for the synthesis of proteins, for making ATP, for the fixation of nitrogen, for the production of methane, and so on. I estimate there are only about 1500 core genes required for the synthesis of all the nanomachines in nature. All of them occur in microbes. That estimate may be a bit conservative—but let's assume that even if it is off tenfold, it means that of the approximately 60 million to 100 million genes in nature, only between 0.0015% and 0.025% of them contain critical information for life. The other 99.98% of genes are related to the function of specific organisms. The vast majority of the 99.98% of the genes are transient—they may evolve to acquire new functions in some groups, disappear in others, or just change in a neutral fashion as the organisms move through time. However, the core genes cannot be lost or significantly altered. Were that to happen, it could spell disaster. Unless a replacement machine evolved in relatively short order, the loss of a core gene potentially could interrupt the flow of key several elements across the planet.

Because the genes encoding for the parts of the core nanomachines are so highly conserved, I refer to them as "frozen metabolic accidents." Although these genes may have evolved for another purpose or under very different environmental circumstances, they have been passed on virtually unchanged from generation to generation of microbes, and from microbe to microbe. They are not necessarily perfect machines, but they work. And nature has evolved several schemes to preserve the genes that encode the core machines, even if those machines aren't perfect.

There is often a misconception about evolution and optimization in nature. The idea is that natural selection, operating over millions and millions of years, will optimize processes critical to an organism's survival and ability to reproduce. Let's examine how that basic idea played out in three nanomachines.

The D1 protein in the reaction center of all oxygen-producing photosynthetic organisms is derived from a nearly identical protein found in purple nonsulfur photosynthetic bacteria that cannot split water to make organic matter. In the absence of oxygen, and only in the absence of oxygen, the purple bacteria are photosynthetic, but they use hydrogen or carbohydrates as the source of their electrons and protons. In these bacteria, the evolutionary ancestor of the D1 protein is highly stable, but in all photosynthetic organisms that produce oxygen, the protein is destroyed after it has processed about 10,000 electrons. "Destroyed" means it not only ceases to function, but it literally starts to fall apart. Effectively that takes about 30 minutes.

What was the solution to this problem? Rather than a new D1 protein being redesigned (evolved) from scratch, an elaborate repair mechanism developed in photosynthetic organisms that split water. The repair system involves identifying the damaged D1, removing the protein from the rest of the machine while it is still in the machinery, and replacing it with a new protein that fits into the hole where the damaged protein was. The situation is something like having to take a set of mechanics with you on every trip in your car, and every 10,000 revolutions of every tire, the mechanics have to lean over, figure out which tire is damaged, and then replace the damaged tire while the car is still moving. In the case of D1, that took a lot of evolutionary monkeying. But it also allowed the old machinery, which was derived from the purple photosynthetic bacteria, to operate under new conditions—in the presence of oxygen.

The damage to D1 is caused by the presence of certain forms of oxygen, forms that are missing electrons or have too many electrons. These so-called reactive oxygen species can cause a lot of damage to proteins, and several enzymes evolved that detoxify them. However, oxygen itself is also highly reactive, especially with nanomachines that contain iron. One of those machines is nitrogenase, which was discussed previously. Like the photosynthetic apparatus, nitrogenase is something of a Rube Goldberg contraption and contains two large proteins that work together to ferry electrons and then protons on to nitrogen gas. In the absence of

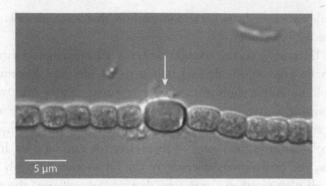

Figure 23. An image of a heterocyst. In some chain-forming species of cyanobacteria (e.g., Fig. 17A), when cells begin to reduce (fix) atmospheric nitrogen gas (N_2) to ammonium (NH_4), they produce a special cell, a heterocyst, in which the oxygen-evolving reaction center (Photosystem II) is lost. The enzyme responsible for nitrogen fixation, nitrogenase, is found exclusively in heterocysts, where it is protected from damage by oxygen. This is one of the earliest examples of cell differentiation in biology. (Courtesy of Arnaud Taton and James Golden)

oxygen, nitrogenase functions pretty well, but in the presence of oxygen, the iron atoms start to "rust," the machinery stops working, and the whole system has to be replaced. One would have thought that after a couple of billion years, that is, since oxygen has been around on Earth, nature would have found an evolutionary path to allow nitrogenase to operate in the presence of oxygen, or perhaps a different type of machine that performs the same function would have evolved. But that didn't happen.

In the case of nitrogenase, the solution was to physically separate the machine from oxygen. In some cases, the cells with the enzyme were confined to anaerobic environments, but in other cases, specialized cells evolved that were slightly less permeable to oxygen than nitrogen (and that is hard to do, because the physical size of the two gases is almost the same). In still other cases, special processes were added that consumed or physically removed the oxygen from the nitrogenase machinery. In no case is the solution perfect. In the oceans today, about 30% of all the nitrogenase is inactivated by oxygen at any one moment in time.

That represents a lot of investment in a junkyard of used parts, which ultimately must be recycled to make new nanomachines.

The last example is even more perplexing. It concerns a very old nanomachine: Rubisco (an acronym for ribulose bisphosphate carboxylase/oxygenase). Rubisco is a protein complex that is responsible for fixing carbon dioxide in all oxygen-producing photosynthetic organisms as well as in a number of other microbes, including many of the chemoautotrophs. It is sometimes said that Rubisco is the most abundant protein on the planet—and for a good reason—although it is responsible for making most of the cell stuff on Earth, it is a pretty inefficient enzyme.

Rubisco isn't all that complex, but it is a big set of proteins: it has two subunits that have to work together. When it works correctly, it takes carbon dioxide, as the gas dissolved in water, and adds it to a five-carbon sugar that has two phosphate "handles" (ribulose bisphosphate) to form two identical, three-carbon molecules. That is, arguably, the most important biochemical reaction on Earth. It is the first step that leads to photosynthetic production of about 99% the organic material upon which the rest of life depends. All animals, including us, are completely dependent on Rubisco for our very existence.

Like the D1 protein and nitrogenase, Rubisco evolved long before oxygen was present in our planet's atmosphere, but it also evolved when carbon dioxide concentrations were many times higher than they are today. Under those conditions, Rubisco works reasonably well. However, in the presence of oxygen, the enzyme often mistakes the gas for carbon dioxide, which is pretty difficult to imagine as the two molecules have different structures. Regardless, if it makes that mistake, it incorporates oxygen to make a worthless product. That happens about 30% of the time in most plants and is a big waste of energy.

To add insult to injury, the carbon-fixing nanomachine is also very, very slow. Each molecule of Rubisco cranks out a product only about five times a second; about 100 times slower than most of the other enzymes in a typical photosynthetic cell. Even the most efficient, most recently evolved Rubiscos are very slow compared with many of the other nanomachines in cells.

One might think that with a slow, inefficient machine and a few hundred million years to redesign it by mutations followed by selection, nature would have evolved a better system. Remarkably, that hasn't happened. While there have been minor improvements, the basic solution has been for cells to make a whole lot of the enzyme. That is a huge investment for a photosynthetic organism. It takes a lot of nitrogen to make Rubisco—nitrogen that could be put to better use to make new cells faster were it not for the inadequacies of the nanomachinery responsible for fixing carbon.

Given the inadequacies of these and many other core machines, one wonders why the machines haven't evolved to be more efficient. Why are the genes encoding these frozen metabolic accidents unable to evolve more efficient nanomachines? The answer appears to be relatively straightforward. In most cases, the nanomachines consist of several components that function as a unit—a literal nanomachine—that physically moves. The movement and orientation of the entire complex of parts is dependent on the individual components. While very small changes in one component may not alter the ability of the nanomachinery to move, large changes in one component without simultaneous changes in others lead to a loss of function. In effect, nature's solution is similar to Microsoft's. When Microsoft first developed an operating system for computers, the software was adequate for the early machines, but as the machines became more complex, Microsoft added more and more code to modify the old software rather than redesign software from scratch. Similarly, rather than redesigning machines from scratch, nature recycles the old machinery and slightly modifies it or develops a set of other components to help it function in a changing environment. Essentially, nature adds more "code" to previously evolved machines.

While the genes for the core nanomachines are highly conserved, many of the remaining 99.98% of the genes that comprise organisms are highly variable. That is, the core machines are found in a very wide range of organisms that often have very distant evolutionary ancestors. For example, in microbes, nitrogenase is found in many groups of bacteria and several groups of Archaea (but not in any known eukaryote). Similarly, Rubisco is found in many organisms that have very little in

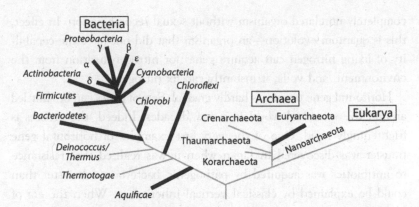

FIGURE 24. A distribution of nitrogenase genes across the tree of life. Note that pattern of the distribution does not follow that of descent from a common ancestor but, rather, is not easily predictable. The genes (and many others) were acquired by horizontal gene transfer between bacteria and between bacteria and archaeans. The genes for nitrogen fixation have never been found in the genomes of eukaryotic cells. (Courtesy of Eric Boyd)

common. A form of Rubisco that is prevalent in bacteria is also found in dinoflagellates, which are eukaryotic algae, but in no other eukaryotes. In fact, the pattern for the distribution for most of the core machines across the tree of life is often unpredictable.

Constructing trees of life that include nitrogenase, Rubisco, and many other core genes clearly reveals that Darwin's model of evolution by descent with variation does not apply. Was Darwin's concept of evolution wrong?

In the era of increasingly faster, cheaper, and better gene sequencing technologies and computers, the complete genomes of thousands of microbes have been analyzed. Inspection of the arrangement of genes in the genome has clearly shown that many genes were not vertically inherited; that is, they were not inherited from a previous generation. This mode of inheritance is called *horizontal* (or *lateral*) gene transfer. Horizontal gene transfer is not a biological curiosity; it is a major mode of evolution in microbes. Simply put, genes that were preadapted via selection in one organism can somehow be transferred to another, completely unrelated organism without sexual recombination. In effect, this is quantum evolution—an organism that did not have the capability of fixing nitrogen can acquire genes for nitrogen fixation from the environment, and voilà, it instantly can fix nitrogen!

Horizontal gene transfer is hardly gradual. Sets of genes can be shuttled around the microbial world in a few decades. Indeed, the process is frighteningly rapid. One of the very first examples of horizontal gene transfer was discovered in Japan when it was realized that resistance to antibiotics was acquired by pathogenic bacteria much faster than could be explained by classical vertical inheritance. When the era of gene sequencing came into its own, it was quickly shown that genes for resistance to many common antibiotics are spread all across the microbial world. It was also observed that many, many other genes are out of place within genomes. Two microbes that, based on sequences of nucleic acids in ribosomes, are thought to be identical, will almost invariably have different arrangements of genes. Rather, many genes appear to have been inserted haphazardly into a genome. In many cases, one or more genes will be inserted between a set of genes that have no apparent relation to the genes in front or behind the inserted genes. The inserted genes are often acquired from a wholly unrelated organism via horizontal gene transfer.

The genes that are transferred were preevolved in other organisms and ferried along, like an unwitting organ transplant to a new recipient who didn't even know s/he was missing an organ. The genes work. Guaranteed. They worked in the organism from which they came for hundreds of thousands (if not millions, and in some cases, billions) of years. They don't have to be monkeyed with to fire them up. If the organism that inadvertently acquired them doesn't need them, they are discarded. If they add to the functionality of the organism, they are used. For microbes, the environment turns out to be a global genomic shopping mall. Preadapted genes are available to any organism that can acquire them, and every organism, including us, has acquired genes via horizontal gene transfer.

How are genes transferred between microbes?

There are three mechanisms known that allow genes to be horizontally transferred, but how they actually work and whether one is more important than the other two remains unclear. The easiest to describe was discovered in the early 1940s by three American biochemists and is called *transformation*. It is outrageously simple—genes (or any DNA) are simply taken up from the environment. A small fraction of the time, the newly acquired genes are incorporated into the new host and are carried along to succeeding generations. While this process works in the laboratory (and the experiments actually formed a strong basis for understanding that nucleic acids rather than proteins contained the information for inheritable traits), it is not clear that there is much free DNA in the real world. Cells don't simply spew out DNA—the cells have to die, and die in such a way that their DNA passes into the environment intact. Which leads us to another possible mechanism for transferring genes horizontally.

The most obvious door-to-door sales representatives of foreign genes are viruses, which come in a variety of shapes and sizes. Many of them look like ultrasmall balls that were designed by Buckminster Fuller; others look like microscopic lunar landers. Regardless of their physical shape, viruses are not alive in a traditional sense. That is, they do not exchange any gases with the environment, have no mechanism to generate energy on their own, and most important, cannot replicate by themselves. They have no ATPases or ribosomes and therefore cannot make proteins or anything else without a host cell. However, they carry genetic information in the form of DNA, or sometimes RNA, packaged in a protein coat. There are extremely large quantities of viruses on Earth. In the upper ocean there are several hundred million viruses in every milliliter of seawater; that is more than ten times all of the bacteria and other microbes together.

The overwhelming majority of the viruses are not well characterized, and in some cases, especially for those that carry RNA, their genetic information changes so rapidly that describing them is a microbiological version of whack-a-mole: the virus you characterized last week is often a different virus this week. If you got last year's flu shot, you probably aren't protected from this year's influenza virus.

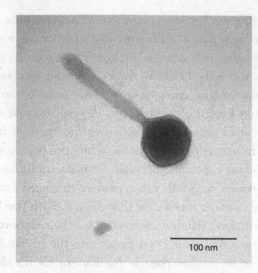

100 nm

FIGURE 25. A micrograph of a marine virus particle. The genetic information of the virus is contained in the head, while the stalk is used to attach to the host cell (e.g., a bacterium). The virus injects its genetic material into the host and uses its machinery to replicate many more viruses. Note that this particle is about one-tenth the size of the smallest cyanobacteria (Fig. 17B). (Courtesy of Jenn Brum and Matthew Sullivan)

Do viruses transfer genes? In principle, yes, but mostly only over short evolutionary distances. Viruses attach to and insert their genetic material into cells, but they tend to have fairly strict requirements for their hosts. They recognize their hosts by specific proteins on the surface of the host's cell, and when they find a suitable host, they can attach to and then transfer their DNA or RNA into the host's cells. The genetic material becomes incorporated into the host and hijacks the host's nanomachines for making proteins and nucleic acids to create new viruses. In some cases, the virus simply is reproduced in the host cell forever—it becomes part of the genome of the host cell. In humans, those types of viruses can be extremely bad news. Two examples of these *nonlytic* (because don't lyse the cell) viruses are HIV and hepatitis C. When they infect a human, they are almost impossible to remove from the genome.

In other cases, however, the newly inserted genetic information allows new viruses to grow inside the host cell until they reach a certain population threshold—then the host cell bursts, releasing the new viruses to the environment. This invasion-of-the-body-snatcher scenario is pretty common in the microbial world—and leads to the death of a lot of microbes. These *lytic* (because they lyse the cell) viruses are also found to infect humans—and, perhaps surprisingly, they are less lethal than the viruses that do not kill cells outright. These types of viruses include those that cause a common cold. Lysis doesn't directly lead to transfer of genes to new hosts, but it does allow the host cell's genetic information to spill out into the environment, where it can be taken up by microbes searching in the genetic wastebasket for leftovers.

The third type of process is called *conjugation*, in which microbes exchange DNA by attaching to each other and forming a bridge between the two cells. This process occurs between closely related microbes, but it not clear how or why it would transfer genes across organisms that are distantly related.

Regardless of the mechanism, horizontal gene transfer makes it very difficult to determine the ancestry of the organisms in deep time, but even more important, it makes the concept of a species in microbes difficult to define, if not irrelevant.

Imagine that you wanted to find out your ancestry. You find where your parents were born, their parents, and so on—but imagine that some thirty or fifty generations ago, genes that digest carbohydrates in seaweed were inserted into the microbial community in the guts of your family because your ancestors ate a lot of sushi. You now are now better adapted to eat seaweed. The microbes in your intestines have new genes acquired from another microbe via horizontal gene transfer. This seemingly absurd scenario, in fact, happens. Microbes in the guts of Japanese have genes that aid in the digestion of seaweed; these genes are not found in the microbes in the guts of Caucasians.

There are a lot of viruses in the ocean that carry the genes for the D1 protein around in their genome, but not because they are evolving to become photosynthetic; rather, the gene for the D1 protein contains instructions for rapid replication. The viruses take advantage of those instructions and use them to make a large quantity of themselves quickly in an infected host cell. But from time to time, copies of the D1 gene from one cyanobacterium are found in a distantly related organism. They presumably got there by viral infection.

Early in the history of Earth, long before there were animals and plants, horizontal gene transfer among microbes was a major mechanism that successfully ferried genes through vast swaths of geological time. The identity of the specific organism is irrelevant, and the scrambling of genes isn't actually critical to life. As long as organisms carry the information that allows energy from the outside world to be converted into a state that is far from thermodynamic equilibrium and the cells can reproduce, life persists.

The scrambling of core genes among many otherwise distantly related microbes helped ensure that the information was retained in some cell somewhere on Earth. Organisms are transient—even disposable—but the 1500 core genes are not. Those core genes of life are transferred like batons in a relay race: organisms carry the genes for vast stretches of geological time and then pass them to new organisms. Individual organisms can go extinct, but as long as they have passed on their core genes to another organism somewhere, those genes will live on.

Horizontal gene transfer probably was important for the early evolution of multicellular organisms, such as plants and animals, but it isn't a principal mode of evolution now. If some of the genes that aid in digestion of sushi were eaten by your great, great, great, great, great, . . . grandparents, assimilated into their genes, and transmitted to their eggs or sperm, you may very well have the genes from the microbes that evolved the gene to digest sushi. But that scenario isn't happening very often. It is prevented by sex.

Sex helped reduce the prevalence of horizontal gene transfer. Genes from other organisms usually don't get into our reproductive cells. Sex helped keep horizontally transferred genes out of germ cells, the cells that make new organisms from sexual recombination. For most microbes, most of the time, sexual recombination is not an option; most of the time they replicate by "simple" cell division, and each new daughter cell is almost always an exact copy of its mother. Sex changed that. Sex scrambles genes from two parental lines. The new cell has a new combination of genes. Although sex allowed greater genetic variation and became the dominant process in the evolution of animals and plants, the process didn't spring up overnight. First there was another, more massive invasion of the body snatchers. The evolution of eukaryotes is the story of horizontal gene transfer on an immense scale—wholesale invasion of one organism by another. Let's see how that happened.

CHAPTER 7

|||||||

Cell Mates

O ne of the strategies nature uses to insure that its intellectual property is resilient in the face of potential massive catastrophic events is to spread the risk across a wide range of microbes. The instructions for nanomachines are spread by means of horizontal gene transfer. Although horizontal gene transfer is the principal mode of evolution in microbes, the process is not totally haphazard. One of the major drivers is ecological—the symbiotic association of microbes to optimize the use of scarce nutrients. That driver has served the evolution of life well.

Microbes do not live in isolation; most of them are *symbionts*, that is, they live together and depend on each other for resources. More specifically, microbes use each other's waste products for sustenance. The use of waste products—also known as the recycling of elements—is one of the basic concepts in ecology, and it has strongly influenced the evolution of microbial nanomachines. It took microbiologists a long time to appreciate the interactions of their subjects on a global scale, but ultimately that appreciation has led to a far better understanding of evolution of life on Earth.

For decades, the approach that microbiologists used to study microbes was to isolate single cells from some environment and try to grow them as pure cultures. These *clones*—colonies of cells established from a single mother cell—were the gold standard, and their use was established by Koch as one of his four principles for proving that a specific organism was responsible for a specific disease. The approach is not without value. Often, very small variations within a population of a single microbial species leads to large changes in the clone's ability to cause disease.

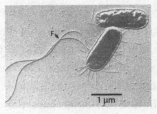

FIGURE 26. An electron micrograph of *Escherichia coli*, probably the most studied microbe in biology. This organism is in human guts, but pathogenic strains (which look identical to nonpathogenic strains) often cause food poisoning in humans. The organism has flagella, which allows it to swim in liquids.

A classic example is that of food poisoning caused by the common bacterium, *Escherichia coli*, which is found in every gut of every single one of us.

E. *coli* is probably the most studied organism in biology. It is very easy to grow, has a wide distribution, and became the model of models for microbial genetics. Small variations in the genes of this organism, which can be transmitted in food, can lead to massive and sometimes very nasty outbreaks of gut infections, and even death, in humans. In this context, an examination of a clone's nutritional requirements, its growth rate, its antibiotic sensitivity or resistance, and so on is extraordinarily important. But with the widespread use of gene sequencing, it quickly became obvious that benign strains of E. *coli* can rapidly become pathogenic and cause massive internal bleeding if ingested. The benign strains acquire pathogenic genes from another strain through horizontal gene transfer via conjugation—a microbial version of sex, which in this case allows a virulent strain to transfer genes that cause disease in humans to a benign strain. Only a small number of genes are required to make E. *coli* pathogenic. The pathogenic strains diverged from the benign strain about four million years ago, but it was very hard to pin down the variations between these two strains in this best-studied microbe until the advent of gene sequencing. If we can't differentiate between two strains of E. *coli* very well without isolating them in pure culture and sequencing their genomes, how are we going to understand microbes in the world around us?

More than 99% of the microbes in the oceans, in soils, on the surfaces of rocks, or even in our own guts that have been identified by gene sequencing have not been isolated and cultured in the laboratory. Many attempts have been made to isolate a myriad of microbes from the oceans, soils, hydrothermal vents on the seafloor, our guts and mouths, and many other environments. Sometimes these attempts have been successful, in that a new microbe has been coaxed to grow in a pure culture; but most have failed. For a long time it was, and often still is, assumed that the lack of our ability to isolate microbes as pure clonal cultures was because scientists simply didn't know the nutrients that these apparently fickle individual organisms require for their growth. How much sugar and what type, which amino acids, and how much salt does each individual microbial species need? The combinations are virtually infinite. In this respect, humans have almost no clue about how microbes function. And so in the laboratory, where the usual objective is to get a lot of microbes to grow as quickly as possible, they are given an abundance of sugars, amino acids, or whatever it takes to coax the organism to grow. The concentration of nutrients in laboratory broths is thousands of times greater than in most real-world situations. With very few exceptions, sugars, amino acids, and the other nutrients are very scarce in nature, and it takes a lot of energy for microbes to acquire them. Understanding how microbes make a living in the real world required a new approach. Microbial ecologists became, in effect, social scientists who study the interactions among microscopic organisms.

To minimize the energy spent acquiring nutrients, microbes in nature tend to form communities in which, for example, a sugar secreted by one organism is consumed by another, while the recipient of the sugar provides amino acids to the others in the community. It turns out that, by and large, microbes, like us, are social organisms. What they lack in complex behavior, they make up for with innovative metabolisms based, to a large extent, on the flexibility of their nanomachines to adapt to changes in their environment.

Microbial communities, or *consortia*, are microscopic jungles in which tens or even hundreds of species of microbes live in a mutual habitat. It should be noted that it is often difficult to strictly define what a microbial "species" is. The traditional definition of the word—that the offspring from sexual recombination is viable— which is testable in animals and plants, usually does not readily apply to microbes. Not only is it difficult to define sex for most microbes, horizontal gene transfer makes defining "species" somewhat specious. Regardless, for the purposes of understanding the function of a microbial consortium, let's consider a microbial "species" within the context of an observable biological function, specifically metabolism. Let us imagine that one species of microbe emits some secretion or gas to the environment, and a second species may use that as a source of energy. The second then emits its own secretions and gases, which can be cycled back to the first species or onward to other species or both. The result is the emergence of a microscopic microbial community that is, in effect, a miniature biological electron marketplace.

The concept of an electron marketplace in microbial consortia is not a metaphor. Microbes within the consortia literally exchange gases and other materials that have an abundance or dearth of electrons. For example, both methane and hydrogen sulfide have an abundance of electrons. These reduced molecules may be produced by several different microbes within a consortium and secreted into the environment. These electron-rich molecules are used by other microbes as sources of energy. Their secretion products—for example, carbon dioxide and sulfate—may be recycled or lost from the community to the external environment. Microbial consortia can be stable for days, decades, or even longer; we simply don't know, but the answer is probably all of the above. However, we do know some of the basic rules of consortia.

One rule within a microbial consortium is that no member may outcompete all the others to exclusion. Should that arrangement be violated, the consortia would collapse, and the "winning" microbe would be at an energetic disadvantage—having to shop for scarce nutrients in distant markets rather than living and dining in the luxury of having locally produced nutrients delivered directly to their microbial doorsteps.

Does that mean that all the bugs "play nice"?

Microbes may be social, but they are also capable of being aggressive and competitive. They often can make molecules that kill other microbes. Indeed, most of the most important antibiotics for combating infectious diseases are made by microbes. But within the context of microbial consortia, these molecules often serve as defenses against invaders, not for killing microbes within the consortia. In other words, to the best of our understanding, there is a gentlemen's agreement that specific bugs with specific functions are allowed into the eating club, whereas others are excluded.

We can easily examine that hypothesis. Humans are born without any microbes in their guts. Very quickly, we acquire microbes from the environment. We get microbes from touching and sucking on our mothers after we are born; we eat raw food; we eat some dirt; we may even scoop some poop. In fact, one of the first microbes to colonize our guts is E. coli, which, one hopes, is a benign strain.

Over time, each of us begins to cultivate a microbial zoo in our guts that is unique—maybe even more unique than our individual DNA sequences. The total number of microbes in each of our guts is about ten times larger than the total number of cells in our bodies. Not only are the microbes in our guts tailored to our personal diet and environment, the composition of the consortium is also extremely important to our personal health. The consortium helps us acquire nutrients from our food by aiding in the breakdown of complex carbohydrates and fats, it helps make vitamins for us, and it helps to keep "bad" microbes from making us sick by preventing their growth. We all know this in some way. Anyone who has traveled to a foreign country, drunk water from the tap, and gotten sick wonders why the indigenous people didn't all die in childhood. In fact, many may, but the survivors have microbes in their guts that can protect them from microbe-borne diseases in the water they drink. You didn't acquire those protective microbes in the food or water back home. If you lived long enough in the foreign country or had been born there, you would have the microbes, or you would have become emaciated, died, or at least not been very successful in reproduction.

Now it is often the case that we will get sick at some time in our lives and a physician will prescribe an antibiotic, or maybe two. We will take the course of antibiotics, of which one side effect is often gastroenteritis; the collateral damage from taking an antibiotic is the death of some of the microbes in our guts. It not only doesn't feel good, it also alters the interactions of the microbes within our gut consortium. It takes some time, sometimes months, for the consortium to come back to the same state as before the antibiotic treatment. In some people, it may not come back even after a year. In other cases, the adjustments are hard to make, and some of us are sensitive to foods we used to consume for quite a while after the course of antibiotic treatment. The personal relationship we have with our gut microbes, which in total compose about two kilograms of our body mass, could be considered a microcosm of what microbes do on a global scale.

Consortia are microscopic representations of the global electron market, but any individual microbial group within a consortium is invariably missing one or more key metabolic pathways to keep the group in energetic balance. For example, one group may be capable of fixing nitrogen—but that function may not be needed if there is a surplus of nitrogen in the consortium. One group may fix carbon—but that element may not be limiting the growth of the consortium. One (but usually more) key reaction is always missing or out of balance. What this means is that the recycling of nutrients and gases within a consortium is never perfect, the consortium is continuously tweaking the electron market to remain viable.

There is always a measureable, net exchange of gases between the environment and a microbial consortium. For example, consortia either consume or produce oxygen, carbon dioxide, methane, sulfur dioxide, hydrogen sulfide, nitrogen, or some other gas. In fact, by following the exchange of gases with the environment, we can often tell what sorts of microbes are in the consortia. In effect, although consortia are relatively self-contained, they always leak gases to the external world. The gaseous waste products are carried away via the atmosphere or oceans, which effectively act as wires, connecting the metabolism of microbes across the globe.

Let's consider this concept on a local, personal level by reexamining our own guts. Without going into intimate detail, it is clear that our, personal, microbial consortia are also not in equilibrium. Most of our gas exchange with the outside world is via our nose and mouth. However, we have another mode of gas exchange, and that mode tells us a great deal about our microbial consortia. Virtually all of the gases coming out of the anaerobic guts of all mammals are oxidized—nitrogen and carbon dioxide being the most prominent. But some are reduced gases, of which the sulfides are the most obvious to our olfactory system. Two other reduced gases, which have no detectible odor, are methane and hydrogen. About half of us have methane-producing microbes in our large intestine, and almost all of us emit hydrogen gas. These two gases are flammable. All the gases produced by microbes in our guts are by-products of metabolism that are not in equilibrium with the local environment. If they were, the gases would be similar to those in our planetary atmosphere, and clearly that is not the case. If the gas mixture in our guts is not at equilibrium with the atmosphere of the planet, then it follows that the ensemble of all the consortia of microbes in the guts of all animals are not in equilibrium with metabolic pathways on the planet. To allow the exchange of electrons among the myriad microbial consortia to work on a global scale, there have to be some global checks and balances, which scientists often call *feedbacks*.

With few exceptions, the change in concentration and composition of gases in the global atmosphere due solely to natural processes are not normally measureable on time scales of centuries. Microbes create a global market for electrons that is stabilized by the integration of the metabolism of hundreds of billions of consortia spread across the planet—from the surface films on lakes to hundreds of meters into the sediments and rocks in the deep sea. The metabolism of Earth is an outcome of a consortium of consortia in which an individual consortium is dispensable, but the machines for all electron transfer reactions are distributed nonrandomly, depending on opportunity and accessibility of resources. Nature's insurance policy is to spread the risk primarily by investing in a global microbial electron hedge fund. The investment is in the potential of nanomachines to operate based on the availability of any molecule in the environment that can serve as either a source of or a sink for electrons.

On a microscopic scale, the organisms within a consortium are living in very close proximity. Under such circumstances, the opportunity for successful horizontal gene transfer is greatly enhanced. Hence, within consortia, gene transfers often allow a distribution of metabolic nano-machines across many groups of microbes, thereby allowing the flows of elements between organisms to be tightly controlled. On a global scale, the operation of these nanomachines has led to a macroscopic engine of life that controls the fluxes of key gases.

Controls are embedded in the chemical signals that are sent from microbe to microbe within the community and that provide information about who is doing what and how many are where. The system of intercellular signaling, called *quorum sensing*, resulted from the evolution of specific molecules that are made and used by microbes to assess their own population density, as well as to signal other microbes about who and where they are. This mode of intercellular communication remains pretty remote to us, although we do know that there are specific molecules sent out by some cells that float around until they attach to specific receptor sites on another microbe's membrane. Much like what the perfume companies would like all men to sense in women or vice versa, the molecules produced by microbes signal to other organisms who and where they are.

Once attached, the molecules work by altering the expression of genes in a cell. Quorum sensing allows consortia to establish a spatial pattern of microbial metabolism that further increases the efficiency of recycling of nutrients. But it can also alter behavior.

At this point, one might reasonably ask, do microbes "behave"? The answer is yes; they don't have brains, but they do have sensory systems, and some of them can be quite sophisticated. They can sense signals from the environment and each other, transfer the signal to a receptor, and trigger a response. Let's examine one example, which led to the discovery of quorum sensing.

Quorum sensing is an example of an emergent property of social interactions of microbes. It was discovered by accident in 1979 by two friends and colleagues, Ken Nealson, then at the Scripps Institute of Oceanography, and the late J. Woodland (Woody) Hastings, at Harvard University. They were interested in how luminescent bacteria that live in the light-emitting organs of some marine fish work. In such organs, the bacteria are found at extremely high densities, upward of 100 billion cells per cubic millimeter. When the microbes from the organs were isolated and grown in pure cultures at low cell densities, they were not luminescent; however, as the cells grew and the population density increased, the colonies started to glow. Nealson and Hastings knew that there is a specific set of genes required for the production of light in the bacteria. These genes are somehow switched off when the cells are grown at low concentrations and switched on when the cells are highly concentrated. The researchers discovered that the signal that switches the genes on is a chemical that the cells secrete, and when its concentration is sufficiently high, the cells literally light up.

Subsequently, many microbiologists have worked on quorum sensing, and although there is still a lot left to learn about the phenomenon, we understand some of the basic principles. It has become clear that microbes use chemical signals to turn on and off various functions within their own population and between populations of other microbes. These chemical signals are harbingers of increased complexity, but they do not necessarily require the evolution of new nanomachines. Microbial communication via chemical signaling is a key mechanism for regulating metabolism among the various groups of organisms within consortia. But something else can also occur.

As in any situation in which many different organisms live in close proximity to each other, there may be unexpected consequences. One of these appears to have happened more than two billion years ago, when one microbe engulfed another and not only retained a subset of the engulfed organism's genes, but also retained the engulfed organism. This wholesale horizontal-gene-transfer process was given the name *endosymbiosis*—a symbiotic association within a cell, or more correctly, a symbiotic association between two cells, one of which is housed within the other.

The original concept can be traced to a publication in 1883 by Andreas Schimper, the German scientist who first described chloroplasts. He observed that chloroplasts in plant cells divided in a manner similar to cyanobacteria and logically thought that the chloroplasts actually were cyanobacteria living within the cells. Schimper's hypothesis was picked up by a Russian botanist, Konstantin Mereschkowski, who studied lichens, which are symbiotic associations of photosynthetic microbes (often cyanobacteria) and fungi. In 1905, Mereschkowski published a paper in

Russian and German, "On the Nature and Origin of Chromatophores in the Plant Kingdom," in which he suggested that chloroplasts were symbionts within plant cells. His work was largely forgotten during the First World War and the ensuing Russian Revolution, not because of those events per se, but because of a sex scandal. Mereschkowski was accused of being a pedophile, and in 1918, he fled to France and then to Switzerland. He continued to write about symbioses but committed suicide in 1921, and his ideas languished in obscurity.

The basic idea that an intracellular body could have at one time been a free-living bacterium that was engulfed by a host cell was elaborated upon in 1927 by an American biologist, Ivan Wallin, who was on the faculty of the medical school at the University of Colorado. He claimed that mitochondria could be grown outside of their host cells. It was later shown that Wallin's samples of mitochondria were actually contaminated by bacteria, so his work was largely discredited.

The endosymbiosis hypotheses got a new push in the early 1960s, when it was discovered that both chloroplasts and mitochondria each contain their own DNA, which is distinctly different from that in the nucleus of the cell, and each contain their own set of ribosomes. Indeed, the matryoshka-doll model of a cell got a big boost, but it also was clear that neither chloroplasts nor mitochondria could replicate outside of their host cells. Moreover, the analysis of the ribosomal RNA sequences by Woese and Fox in both chloroplasts and mitochondria revealed that both organelles are descended from bacteria. That analysis clearly proved that Schimper's and Wallin's basic hypotheses were correct: chloroplasts are related to cyanobacteria, and mitochondria are related to another set of bacteria, members of which, interestingly are anaerobic photosynthetic organisms.

The notion of endosymbiosis finally garnered respect and widespread acceptance in 1967 when Lynn Margulis, an American biologist, wrote a paper resurrecting Mereschkowski's hypothesis, not with new data but rather as a theoretical problem. She went on to argue for the concept in a series of papers and several books. Margulis was an exceptionally articulate scientist and good friend. She spent most of her illustrious career extolling the concept of endosymbiosis as a driving force of the evolution of life on Earth. She was partly correct.

The phenomenon of endosymbiosis is relatively common, but it very rarely leads to the establishment of a new organelle. In fact, the only two organelles we can absolutely be sure were inherited via this route are the mitochondrion and the chloroplast, but the events leading to the incorporation of these two organisms in a host cell altered the course of evolution. Were it not for endosymbiosis, we would not exist. In both cases, the process began in the oceans, long before there was evidence of any significant life on land, and in both cases, chemical signaling became critically important.

The evolutionary history of eukaryotes is not fully resolved. It appears, though, that the microbe that served as the host cell was an archaea, which was similar to the organisms in our guts that produce methane. In one scenario, the organism it ingested was closely related to living purple nonsulfur photosynthetic bacteria. The latter are more ancient than cyanobacteria and can use light energy for photosynthesis only when there is no oxygen in their environment. Under such conditions, they use the energy of light to move electrons around in a closed cycle and build up a gradient in protons across a membrane. The protons then are allowed to flow through the coupling factor to form ATP. This is exactly the same nanomachine we have discussed earlier.

In the presence of oxygen, however, the electrical circuit is inhibited, and the cells lose their capacity to synthesize the pigments that absorb light. To survive, they "rewire" their internal electronic circuits and allow oxygen to become an acceptor of hydrogen that comes from organic matter. The same bacterium that is a photosynthetic Dr. Jekyll during the day under anaerobic conditions can become a respiratory Mr. Hyde under aerobic conditions. During the day it can use solar energy to make itself a net contributor of organic matter to the microbial world, but only if there is no oxygen present. If there is oxygen around, the bacterium transforms itself into a consumer of organic matter and uses the energy in organic molecules to grow. In other words, in the presence of oxygen, the nonsulfur bacteria respire, just like us and all other animals. Animals have retained intracellular Mr. Hydes—the mitochondria.

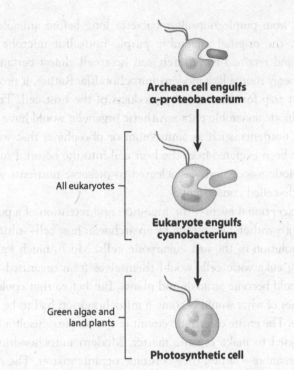

Archean cell engulfs
α-proteobacterium

All eukaryotes ⎨

Eukaryote engulfs
cyanobacterium

Green algae and
land plants ⎨

Photosynthetic cell

FIGURE 27. A schematic showing the two basic endosymbiotic events that led to the formation of eukaryotic cells. In the first event, the host cell (an archaean), engulfed a purple nonsulfur bacterium, which possibly was photosynthetic. The bacterium would evolve much later to become a mitochondrion. In the second event, a cell containing the protomitochondrion engulfed a cyanobacterium. The cyanobacterium would later evolve to become a chloroplast. These two primary symbiotic events are the basis of the evolution of microscopic organisms, such as green algae (Fig. 9), that were prevalent in the oceans long before the evolution of animals and plants.

How did the ingested anaerobic photosynthetic bacterium ultimately become an oxygen-consuming mitochondrion? The nanomachines in the purple photosynthetic bacteria are exactly the same nanomachines we use to generate energy in every cell in our bodies—and that's not coincidental; it is causal. Our power supplies, our mitochondria, were inherited from purple nonsulfur bacteria long before animals evolved. However, the original anaerobic purple nonsulfur microbe that was ingested and retained by an archaean host cell almost certainly wasn't a huge energy source like modern mitochondria. Rather, it probably was a nutrient trap for the excretory products of the host cell. That is, the endosymbiotic anaerobic photosynthetic organelle would have been able to utilize nutrients, such as ammonium or phosphate, that would have otherwise been excreted from the host cell into the ocean. I suggest that the symbiotic association was selected to preserve nutrients within the new, single-celled consortium.

This exceptional event—the ingestion and retention of a purple non-sulfur photosynthetic bacterium by an archaean host cell—ultimately led to the evolution of the first eukaryotic cells. Much, much later, single, free-living eukaryotic cells would themselves form organized consortia, which would become animals and plants. But before that could happen, the engines of what would become a mitochondrion had to be set to run in reverse. The entire electronic circuit of the purple nonsulfur bacterium was designed to make organic matter. Modern mitochondrion do not do that anymore—rather, they consume organic matter. The reversal of the electronic circuit required oxygen, but neither the purple nonsulfur photosynthetic bacteria nor the host cell could make oxygen. That division of labor required another set of skills. But to make this arrangement work for both the host and the newly engulfed cell, the two partners would have to communicate with each other.

In the acquisition of the anaerobic purple photosynthetic bacterium, the host cell had to quickly obtain control over the intracellular organism. Imagine that the intracellular organism could grow even slightly faster than the host. Over a few generations, the intracellular organism would outgrow the host and the host cell would die. Imagine the reverse: that the newly acquired intracellular organism grew slower than the host. The host would then be forced to grow slower and perhaps not be as competitive in acquiring nutrients as its unencumbered relatives that never acquired an intracellular organism. Controlling the newly acquired intracellular organism involved the transfer of key genes from the intracellular organism to the host cell and the loss of many more genes in the intracellular organism. The new, now eukaryotic, cell was now a consolidated microbial consortium in which the host cell effectively enslaved its intracellular, endosymbiotic partner. Over time, the intracellular organism lost so many genes that it could no longer replicate outside of the host; however, it retained some genes for key nanomachines for energy production as well as the ability to make some proteins. Now there were two protein factories in a single cell.

Making sure that one protein factory did not outgrow the other took some doing at the beginning, before the gene transfers and losses occurred. It required chemical signaling between the two cells, a process that is still not well understood. The chemical signals are sent from the mitochondrion to the host cell's nucleus while another operates in reverse. Mitochondria ultimately became very sophisticated. They can turn on and off genes in the nucleus of the host, amplify specific pathways, and alter the host's behavior. This signaling system was given the unfortunate name of *retrograde signaling*, but in essence it is very similar to quorum sensing between two cells that share a single room—cell mates, as it were. It was the first step toward the evolution of cooperation of many such cells that function as a single unit. But before that would happen, a second endosymbiotic event occurred.

In the second major endosymbiotic event, an anaerobic cell that already contained a photosynthetic purple bacterium (the protomitochondrion) took in another lodger. This time it was an oxygen-producing cyanobacterium. This three-body arrangement probably happened many times, because almost certainly most of the attempts resulted in the death of the anaerobic purple photosynthetic bacterium. In its evolutionary history, the purple photosynthetic bacterium had almost certainly never been exposed to significant amounts of oxygen, let alone to a veritable continual gush of the gas when the Sun shined. Let's follow the logic of this miniature microbial zoo and see how it came about.

The waste product of the cyanobacterium, oxygen, had to be used by one of the two other partners, who already had a cozy arrangement. It wasn't in the purple photosynthetic bacterium's interest to take in the new boarder, but now the purple bacterium was faced with having to be potentially poisoned by the oxygen produced by the cyanobacterium in the same host cell. The host was certainly promiscuous, but why would it try to kill its first endosymbiont, which was doing a good job of recycling nutrients? To avoid death and potential extinction, the purple photosynthetic bacterium had to evolve to use the oxygen in some fashion. It found that oxygen, a good electron acceptor, could readily accept electrons from organic matter, but making that process work required the evolution of another nanomachine, one that could transfer electrons and protons to oxygen. The new nanomachine, cytochrome c oxidase, is extremely complex, and components of it predate the production of oxygen by cyanobacteria. Its ancient parts were recruited to form a redesigned complex nanomachine by salvaging and rearranging components from other, simpler nanomachines that are found in both bacteria and archaea. Cytochrome c oxidase almost certainly did not originally evolve to put electrons on oxygen; it probably evolved to remove oxygen from a cell. The modern incarnation of cytochrome c oxidase contains up to 13 protein subunits and uses copper to help in carrying out its chemical reactions. Once this nanomachine evolved, the world would never be the same.

Oxygen allowed cells to become truly supercharged. The use of the electrical field across the mitochondrial membrane allowed 36 ATPs to be made from one molecule of glucose. Cells could now power little motors that allowed hairlike structures, flagella, to rotate, and in so doing the cells could become highly motile. They could develop new metabolic pathways, taking advantage of oxygen and energy to make complex lipids, such as cholesterol, and many other, more complex molecules. The organisms that were the acquisitioners and the organisms that were acquired would come to be permanent cell mates.

The new cell mates, in a mutual prison of their own making, had potential advantages for all the participants, but for this machine to work, the cell mates had to cooperate with each other. In the new arrangement, there were now *three* sets of genetic information in a single cell: the host had a set, the protomitochondrion had a set, and the newly acquired cyanobacterium, a nascent chloroplast, had a set. To get all of these sets to work together so that one of the endosymbionts didn't outgrow the host and the host didn't outgrow the endosymbionts required some alterations and signals.

One of first alterations was the massive loss of genes from the newly acquired cyanobacterium, just as we saw above with the acquisition of the purple photosynthetic bacterium. The cyanobacterium retained some of its genes to make some key proteins, especially ones that form the nanomachinery of the photosynthetic reaction centers, but many of the genes that enabled it to grow outside of the host were simply discarded or transferred to the host cell.

The two endosymbiotic events that formed the basis for eukaryotic cells are examples of wholesale horizontal gene transfers that would endow the new photosynthetic cell with properties that it would never otherwise have had. The origin of nascent chloroplasts in a cell that contained a protomitochondrion would allow for the evolution of many new forms, from individual algae to massive trees. But regardless of form of the body, all eukaryotic photosynthetic organisms use exactly the same ancient nanomachines to generate energy, to make proteins, and to generate new cells.

Ultimately, these new organisms would become increasingly complex and successful. Indeed, following the Great Oxidation Event, the fossilized bodies of eukaryotic cells became increasingly abundant. The research and development phase of life's core nanomachines essentially ended with the evolution of eukaryotic cells.

The rest of evolutionary history was concerned with the body plan; that is, in what body form the nanomachines would be housed. Eukaryotic cells could themselves form consortia and acquire new shapes. They could swim faster and longer than their prokaryotic cousins, which they now consumed for nutrition. But the new eukaryotic cells also evolved novel, more sophisticated communication systems. These sensing systems are myriad chemicals that facilitate intra- and intercellular signaling, an elaboration of quorum sensing. Those communication systems would, over the next 1.5 billion years, evolve into complex integrated multicellular consortia—animals and, later, plants.

Let's now see how and why the nanomachines, derived during 2.5 billion years in microbes, are maintained in the macroscopic consortia of eukaryotic cells, the animals and plants that were so familiar to Darwin as well as to all of us.

CHAPTER 8

||||||||

Supersizing in Wonderland

Why and how did microbes become the organized macroscopic organisms—the animals and plants—that are so familiar to our everyday experience? That evolutionary transformation would seemingly have huge costs. Animals and plants have much slower reproductive rates and a much more limited metabolic repertoire, and they are far less adaptable to changes in environmental conditions than microbes. However, these apparent disadvantages did not preclude the evolution of large, multicellular organisms. Let's examine the evolution of complex, or "higher," organisms and how they were assembled from the smaller building blocks that evolved three billion years earlier in microbes.

The timing of the rise of animals and plants relies on two independent lines of evidence. The first is physical fossils. Fossils of single-celled eukaryotic organisms called *acritarchs* (a term derived from the Greek meaning "of confused origin") became relatively abundant between about 1.8 and 1.5 billion years ago. They had cell walls composed of molecules similar to cellulose, and spines and other external features that are consistent with the resting spores of some extant single-celled eukaryotes such as dinoflagellates. While some acritarchs may have formed multicellular colonies, there is no clear evidence for true multicellular animals or plants until much later.

The appearance of multicellular animals in the fossil record appears as if out of nowhere. Darwin understood that the appearance of many animal fossils in what were then the deepest (and therefore oldest) rocks in Wales, the Cambrian sequence, was problematic from a perspective of evolution, but he had no idea how to reconcile the issue.

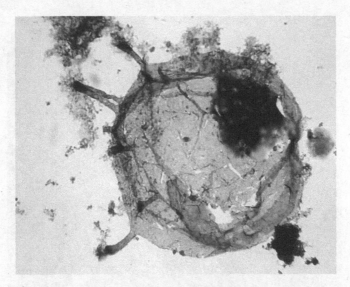

FIGURE 28. A fossil acritarch (*Tappania plana*). This and other members are now extinct but they were the forerunners of modern eukaryotic phytoplankton. This fossil was found in northern Australia and dates to between 1.4 and 1.5 billion years ago. It was quite large; this cell is approximately 110 microns across its diameter. (Courtesy of Andrew Knoll)

In 1868, a Scottish geologist, Alexander Murray, discovered a new fossil in Newfoundland that lay below the Cambrian sequences. The fossil was clearly multicellular, but he had no real idea what it was. It was largely dismissed by paleontologists as an artifact. It wasn't until 1957 that a series of fossils, found in the Ediacara Hills in Western Australia, would be accepted as evidence of Precambrian animal life. Fossils from this period, called the Ediacaran, have subsequently been found in several places around the around the world, including the White Sea in Russia and Mistaken Point in Newfoundland, the area that Murray had described a century earlier.

The earliest animal fossils date to about 580 million years ago. They appear to have evolved after the last global ("snowball") glaciation. The preserved Ediacaran animals were all of marine origin and appear to have been soft-bodied—that is, they did not form shells or skeletons or any

FIGURE 29. A fossil of *Dickinsonia*, an extinct animal found in the Ediacara Hills of South Australia. This and other Ediacaran fossils are the oldest fossilized animals and evolved approximately 600 million years ago in the oceans. (Courtesy of Jere Lipps)

biominerals or hard parts that we can discern. They existed for about 90 million years. The Ediacaran Period ended 543 million years ago and was the first visible extinction of animals in the fossil record.

In 1909, an American geologist, Charles Walcott, from the Smithsonian Institution accidentally discovered a large sequence of marine fossils in the Rocky Mountains in southeastern British Columbia. He ultimately collected approximately 65,000 fossils from that area. More than fifty years later, work by Harry Whittington and two graduate students clearly revealed that this sequence of vertical rocks, the Burgess Shale, contained organisms representing every existing body plan, including early clam-like creatures, segmented worms, and extinct primitive organisms with primitive structures reminiscent of backbones. The Burgess Shale, which formed approximately 505 million years ago, contains an extremely diverse array of fossils. For many years it was debated whether the apparent Cambrian "explosion"—that is, the apparently rapid and extraordinary evolutionary changes in animal body plans recorded in the fossil record of animal life was an artifact of fossil preservation or a true period in animal diversification. It is likely that some animals from the Ediacaran escaped the extinction of 542 million years ago and became seeds of animal life in the Cambrian, but the founder species remain to be identified.

The second line of evidence is less direct. It is based on the concept that the rate of mutations within specific genes, portions of genes, or groups of genes can be determined. If one knows the rate of mutations, then by counting the number of mutations from extant members of a group of organisms, one can infer the rate of evolution of the group. These *molecular clock* models can be used to extrapolate back to the origins of organisms. The more recent models take into account variations in mutation rates and are potentially more accurate than earlier ones. Whenever possible, molecular clock models are calibrated with physical fossils, but inevitably, the further back in time one tries to extrapolate a particular model, the more uncertain the model becomes. Models based on molecular clocks almost always predict origins of organisms that are earlier than the evidence based on the first appearance of the physical fossils in the rock record.

Using a molecular clock model calibrated with fossils, a group of scientists led by one of the best invertebrate paleontologists, Doug Erwin, from the Smithsonian Museum in Washington, D.C., dated the rise of animals to about 700 million years ago—at the beginning of the Ediacaran. But that is not the most significant inference. More important, Erwin and colleagues also made a convincing case supporting the rapid evolution of animals. That is, the Cambrian explosion appears to have been a real period of the evolution of many new animal body plans. Although the timing of the rise of animals is relatively well constrained, the evolutionary innovations responsible for the phenomenon are not well understood.

In thinking about why animals evolved at all, I have often come to a very simple hypothesis. Multicellularity was a strategy for ecological success in environments that contain few food particles. Simply put, starvation was a driver of evolutionary selection. The energetics of single-celled organisms living in an ocean is hard for us to imagine. In a famous and wonderful essay written in celebration of the great theoretical physicist Victor Weisskopf, his colleague Edward Purcell described in a captivating little essay, "Life at Low Reynolds Number," how microbes experience life in a fluid. It turns out that for a microscopic organism, water is a relatively viscous fluid. It takes a lot of energy to move in viscous fluids. The analogy Purcell made was that a human sperm cell swimming in water experiences the fluid as that of a full-sized human swimming in molasses. We would only be able to move a few meters per week. If cells could work together in unison, they would be far more efficient in overcoming the physical barriers imposed by the viscosity of fluid in which they live.

To form a multicellular animal, cells had to evolve four basic traits. They needed a shared power supply. They had to adhere to each other in precise way. They had to share functions communally for the organism, rather than only for themselves. And they had to reproduce that template again and again and again. These four traits had to function together, like a choreographed theater performance. If a multicellular organism failed in any one of the four traits, it would become extinct.

The power-supply issue was critical. With a very few exceptions, animals require oxygen to extract energy from their food. In single-celled eukaryotes, oxygen reaches the power-generating system, the mitochondria, by diffusion, a process in which molecules, which are always moving randomly because of thermal energy, move to where the concentration of oxygen is lower. As oxygen is consumed within mitochondria, the organelles keep the concentration low in that part of the cell, and oxygen then moves from the outside world, which 1.8 billion years ago was the ocean, into the cell.

Diffusion works reasonably well for getting oxygen to single-celled organisms. But if the single cells start to get large and the oxygen concentration isn't very high, the cell won't get enough oxygen, and it won't grow very well. This problem is really exacerbated when cells form colonies and start to become multicellular.

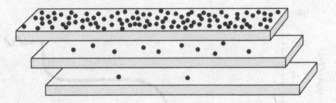

FIGURE 30. The problem of diffusion of oxygen in multicellular animals. Without some circulatory system, oxygen can only be supplied to cells via diffusion. If an animal lives on the seafloor, the only source of oxygen is from the waters above. The oxygen reaching the first layer of cells is depleted by respiration, the second layer receives far less oxygen than the first, and so on. The diffusion of oxygen almost certainly contributed to the evolutionary selection of thin animals in the early Ediacaran Period.

Imagine that an organism is a flat plane, like a paper napkin, living on a surface, such as a rock or a muddy sediment. Let's assume that, like a folded napkin, the organism is made up of layers, but instead of thin layers of paper, they are composed of layers of respiring cells, like the fossil animals in the Ediacaran. As oxygen diffuses into the top layer, 90% is consumed by the cells comprising that layer, leaving only 10% for the next layer of cells. The next layer consumes 90% of that remaining 10%, leaving less than 1% for the third layer. Clearly the cells at the bottom are going to be starved for oxygen and won't function very well.

The situation could be helped if the initial concentration of oxygen was high and if the cells were organized into a shape that allowed oxygen to come from other sides, or if the cells developed a system that efficiently distributed oxygen. All of these solutions ultimately evolved, but the initial condition required a significant boost in the oxygen concentration of Earth's atmosphere.

The burial of organic matter in the ocean's sediments, and the attendant rise of oxygen in Earth's atmosphere, was dramatically hastened with the evolution of phytoplankton. Unlike their prokaryotic ancestors, which barely sank in the oceans because they were so small (the viscosity of water helped keep them suspended), eukaryotic phytoplankton could sink rapidly. Their evolution and subsequent death and burial in the sediments

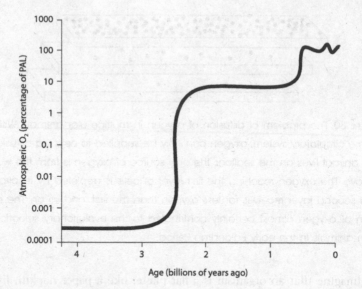

Figure 31. An illustration of our present-day reconstruction of oxygen over geological time. Note that the scale for oxygen is logarithmic. Oxygen concentrations during the first half of Earth's history were vanishingly low, on the order of 0.0001% of the present atmospheric level (PAL). The concentration may have risen to approximately 1% of PAL during the Great Oxidation Event, 2.4 billion years ago, and then rose again to approximately 10% during the Ediacaran and Cambrian Periods, about 600 to 500 million years ago. Over the past 500 million years, oxygen concentrations have remained relatively high and relatively stable, varying between approximately 50 and 150% of the present value.

of the ancient oceans led to a long-term sequestration of organic matter and, as a consequence, helped give a boost in Earth's oxygen concentration (see chapter 5). The boost in atmospheric oxygen occurred about 700 million years ago, approximately 1.7 billion years after the Great Oxidation Event. The second rise in oxygen almost certainly was critical for the evolution of animals.

No one knows with certainty what the concentration of oxygen was when animals evolved, but best-guess reconstructions are that it was somewhere between 1% and 5% of the atmospheric volume. Today it is 21%. It is somewhat ironic that the death and burial of eukaryotic phytoplankton accelerated the rise of an oxygenated atmosphere, which helped lead to the evolution of multicellular animals that would feed on the phytoplankton.

With the rise in the concentration of oxygen, single-celled eukaryotes could aggregate into colonies because diffusion was less problematic. But aggregation required some sort of cell-to-cell adhesion, an intercellular "glue"—the second trait that was critical for multicellular animal evolution. The adhesive role was provided by a set of two types of proteins, the collagens and the integrins, which would become ubiquitous in all animals. These two proteins act like flexible epoxies—they cement cells together and also bind many cell products, such as teeth, bones, and shells. There are many types of collagens but all are characterized by three parallel helices, which are similar to microscopic screws. Ancestral forms are found in prokaryotes. We all know collagens: as dried proteins mixed with flavoring and sweeteners, they are sold as gelatin desserts. The collagens dock to integrins, which are proteins bound to cell membranes in animals. These aren't the only adhesive agents, but they are the most important. In animals, collagens can account for up to 25% of all the protein in the organism.

Forms of both collagens and integrins appeared early in the evolution of animals. They occur in sponges, the oldest species of animals, and hold the cells in specific positions and orientations. As animals continued to evolve, the molecular glues became increasingly important in permitting new and more complex body plans.

The third trait, the diversification of cell functions, is one of the most interesting in the biology of animals and plants. Even the simplest animals and plants comprise several different kinds of cells. In animals there are various kinds of nerve cells, skin cells, digestive cells, and so on. In plants there are various kinds of leaf cells, root cells, and shoot cells. All the various cells in the adult organism came from a single cell, a fertilized egg. Regardless of what the cells do in the adult organism, in every cell that retained a nucleus, the genetic material in each of the cells is identical. That is why we can take cells from our saliva, skin, bone, liver, or lungs and analyze our own genome. But each of these types of cells has come to perform a different set of functions—and those functions are encoded in the genes of each organism. The process of becoming a specialized cell within a consortium is called *differentiation*. In animals, cells that are not yet fated to be a specific cell type are called *stem cells*—cells that can be coaxed into being one or another of many types: a nerve cell or a liver cell, etc. But where did all these different kinds of cells in multicellular organisms come from?

Both animals and plants borrowed and elaborated on a theme that evolved much earlier: in microbes. In cyanobacteria that form colonies, there are some cells that lose their photosynthetic ability and become specialized at fixing nitrogen. The new type of cell is bigger, has a thicker cell wall, and is the only cell type within the colony that can fix nitrogen to form ammonium. Also, it cannot be coaxed back to becoming photosynthetic—even though it retains all the genes to do so.

There are several other examples of differentiation. Many single-celled eukaryotes can undergo some form of *genetic recombination*, and in so doing, transform their cells from one form to another. Genetic recombination is a fancier term for sex—when two cells, each with half the genetic complement of the parent, combine genetic information to form a new cell that replicates itself. In single-celled eukaryotes, the germ cells often look completely different from the parent. Indeed, the origins of sex go far back in evolution; they are found in modern eukaryotic algae. The "spores," or germ cells, have half the number of chromosomes— individual segments of genetic information within each cell's nucleus—of the parent cell and often have very different shapes.

Cellular differentiation became a hallmark of both animal and plant evolution. As multicellular organisms develop, specific functions are acquired by specific cells. In lower animals and most plants, the organism can be replicated without sexual recombination simply by taking a piece of the organism and growing it with an energy and nutrient source. In such a case, the cells retain the flexibility to acquire a new function. However, in the evolution of increasingly more complex animals, this flexibility became lost, and the only path to a new organism would be via sexual recombination, the fourth trait.

Sex leads to the formation of a fertilized single cell, a zygote, which differentiates into new types of cells as it divides and develops into an embryo. The informational system for development and organization of cells in both animals and plants became highly complex, but the basic tool kit was acquired from their single-celled ancestors and is analogous to quorum sensing in microbial communities.

In animals a set of molecules evolved that direct the transcription of genes in cells. These transcription factors, which became very sophisticated, organize the developing animal along an axis and direct cell division and function. For example, in animals a set of homeobox (or in scientific vernacular, *Hox*) genes turn on and off hundreds of genes during the development of the embryo; transcription factors, like the *Hox* genes, are often incredibly conserved. They were first discovered in 1984 in the fruit fly, *Drosophila*, but it was subsequently recognized that similar genes are spread across the animal kingdom, from jellyfish to humans.

A completely different set of transcription factors evolved in plants. One of these is the MADS-box genes, which organizes the development of reproductive structures. There are yet others involved in development of roots and shoots in the early germination of seeds. That animals and plants have different kinds of transcription factors, both of which are universally distributed in their respective kingdoms, indicates that the molecules responsible for controlling the body plans of these two groups of macroscopic organisms evolved after they diverged from their last common ancestor. Because both plants and animals appear to share exactly the same mitochondrion, it is unlikely that the origin of animals was a photosynthetic protist that lost a plastid. This leads us back to the evolutionary selection pressures that gave rise to animals in the first place.

The oldest fossils, from the Ediacaran Period, are not clearly related to any modern animal forms, but molecular evidence suggests that sponges, which are preserved in the fossil record from the Cambrian, are the oldest extant animal phylum. (In this context, a phylum is simply a group of animals and plants that share common body plans. Sponges are in the phylum Porifera. "Porifera" means "bearing pores.") The architecture of modern sponges is relatively simple. These organisms are essentially a scaffold of millions of pores through which water can flow. Sponges are a giant consortium of eukaryotic cells. Their architecture and feeding strategy provides a clue as to how and why animals originally evolved.

And here is where Purcell's vision of life for a single small cell in a viscous fluid, like water, is informative.

There are cells in sponges that appear to be closely related to a group of extant single-celled, flagellated organisms, the choanoflagellates. Choanoflagellates have a small collar composed of microvilli, which are small protrusions of the cell membrane. These organisms use their flagella (a word derived from the Latin for "whip") to move water across their collars, where the microvilli trap bacteria and other small organic particles, allowing the cell to ingest them. The flagellum itself is an ancient nanomachine and is found in prokaryotes as well as eukaryotes, although the structures of the flagella are different between the two groups. In eukaryotes, such as choanoflagellates, flagella are composed of nine doublet strands of a protein, dynein, that surround a core doublet strand of the same molecule. Dynein is a molecular motor: one strand hydrolyzes ATP, and in the process, it bends and slides relative to its neighboring strand, somewhat like moving a strand of rope (the dynein) by moving one hand (the motor) toward the other, grabbing the handle, and sliding the other hand down to grab it again, repeating this process over and over. The result is that the flagella whips back and forth, pushing water. This type of flagellum arose in single-celled eukaryotes and is used for propulsion through water and for feeding, in which a flow of particles is directed to the cell. This fundamental nanomachine would become responsible for a host of processes in animals, from the locomotion of sperm to the digestion of food in guts. While most members of the family of choanoflagellates are free-living, single-celled organisms, a few species can form colonies. Although colonial forms of single-celled eukaryotes are not uncommon, some species of choanoflagellates have genes that allow them to adhere to each other in a very precise fashion.

In 1841, nineteen years before the publication of *The Origin of Species*, a French biologist, Felix Dujardin, noted similarities between choanoflagellates and the morphology of cells that line the interior of sponges. He called those cells *choanocytes*. In sponges, choanocytes beat their flagella in a coordinated fashion, moving tens of liters of water through the interior of a sponge every day. In the interior of sponges, the choanocyctes filter out bacteria and organic particles from the passing water,

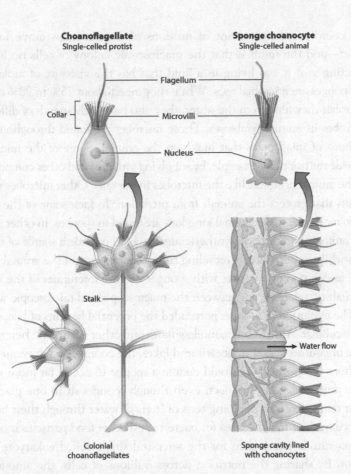

Choanoflagellate
Single-celled protist

Sponge choanocyte
Single-celled animal

Flagellum

Collar

Microvilli

Nucleus

Stalk

Water flow

Colonial
choanoflagellates

Sponge cavity lined
with choanocytes

Figure 32. Drawings of colonial choanoflagellates (left), showing the flagella they use to push bacteria and other particles into the collar, where they are ingested, and the strikingly similar types of cells, the choanocytes, found in sponges (right).

using their flagella to capture and ingest the material for the colony. The movement of the flagella is synchronized to create a unidirectional flow of water through the animal, similar to the way rowers coordinated their strokes on triremes to move the latter through the water. Amazingly, however, sponges don't have a nervous system. It is unclear how the individual choanocytes communicate with each other or what signal may be responsible for the synchronization of millions of flagella. Regardless, the coordinated movement of millions of flagella helps move lots of water—and the result is that the macroscopic colony of cells no longer is acting as if it was living in a fluid that has the viscosity of molasses.

Sponges are microbial zoos. While they ingest about 75% to 90% of the microbes they filter from the water, they also harbor thousands of different microbes in mutual symbioses. Those microbes are found throughout the millions of small pores that make up the animal. Some of the microbes provide nutrition, for example, by supplying vitamins and other compounds to the animal host, just like the microbes in our guts. Other microbes make toxins that protect the animals from predation. In fact, some of the most toxic molecules in the animal kingdom are found in sponges. In other cases, the animals harbor photosynthetic algae, which provide a source of nutrition while simultaneously recycling the waste products of the animal host. The association of microbes with sponges was the forerunner of the wider mutualistic relationships between the macroscopic and microscopic world.

The evolution of sponges portended the potential benefits of becoming multicellular. Although choanoflagellates and other eukaryotic heterotrophic organisms persist in oceans and lakes, the coordinated movement of millions of choanocytes would enable a sponge to access far more water than any single cell. In effect, even though sponges sit in one place for their entire lives, by pumping tens of liters of water through their bodies each day, their foraging area for bacteria and other food particles is orders of magnitude larger than for the ancestral single-celled eukaryote that swam. By sharing the nutrition across millions of cells, the amount of effort expended to acquire food per cell is greatly reduced. Moreover, by having such a flow of water through the organism, the supply of oxygen is sufficient to maintain a high metabolic rate. To top it off, by harboring both nutritional and toxic microbes, sponges are more self-sufficient and less exposed to predation. Cellular networking has its benefits.

The evolution of body plans in animals had been one of the cornerstones of evolution, even before Darwin's time. The idea that an animal that makes a shell, like a clam, is fundamentally different from an animal that has a backbone, like a snake, a bird, or a human being, is obvious at a macroscopic scale. In such a sense, a motorcycle, an automobile, an eighteen-wheel truck, an ocean liner, and a jet airplane are different body plans, but they all contain engines that require a source of energy and all use the same fuels. These human-made machines were invented within a 150-year period—and their evolution, while retrospectively incredibly rapid, is based on the same basic principles of using common machinery to propel differently shaped vehicles. A similar principle holds in the evolution of animals.

The core nanomachines—the coupling factors, the photosynthetic reaction centers, the cytochromes and electron carriers—are responsible for life in all plants and animals and evolved in microbes billions of years earlier. The machinery was appropriated in many body forms first in animals. Animals are a small, relatively irrelevant branch on the tree of life and are like the many versions of motorcycles, cars, and trucks that use the same basic machinery to move. In fact, the metabolic machinery in animals and plants is far less diverse than it was in their microbial ancestors; animals cannot access many of the fuels that were (and still are) available to those microbes. But animals did acquire other novel processes that set them apart from their microbial ancestors.

The novel processes were significant, and while it is not especially critical to enumerate them all, I would like to focus on a few key innovations that allowed animals to become so successful. Some of the more essential processes include long-range motility and sensory systems, and the formation of neural systems and brains. In each case, these systems have microbial origins or analogues; animals modified preexisting genes and did not have to start with new ones.

Motility is one of the earliest innovations in animal evolution. Although sponges are for the most part not motile, their close relatives, the comb jellies, swim. These little animals look like miniature transparent footballs but have eight rows of cells with very large numbers of flagella-like structures, *cilia*, that run along their outside surface. The cilia all beat in unison to create a wave along the outside surface of the animal that propels it through the water. In some respects, the design of this propulsion system is analogous to an inside-out sponge. This system, which was adapted from single-celled organisms, is not very efficient and was abandoned as organisms became bigger. However, it worked well enough to overcome the small-scale viscosity problem experienced by all single-celled organisms in water; comb jellies are the largest organisms to use this system for locomotion. With the evolution of cnidarians, such as jellyfish, propulsion was based on creating a jet of water, which is forced out through their mouth openings.

A miniature football doesn't have great hydrodynamic properties. Navies of the world know this: submarines, which, in effect, are elongated footballs, require a lot of energy to move through water. With the evolution of bilaterally symmetrical animals, such as worms, insects, fish, reptiles, birds, and us, a large fraction of cells developed to become muscles—which are controlled by other cells, nerves—that act in a coordinated way to move the animal very efficiently in water or air. The evolution of all these systems required a set of molecular motors, a role that was filled by a set of proteins called *myosins*, which use ATP to "walk" along another protein, *actin*. For a long time it was thought that the genes that encode the myosin proteins were restricted to animals, especially bilateral symmetrical ones. However, as more and more genetic sequences became available, it became clear that not only did comb jellies and jellyfish contain myosins, but also that the genes were derived from single-celled eukaryotes, especially the choanoflagellates. Animals essentially salvaged and reused genes that had evolved hundreds of millions of years earlier. Millions of years later, the machines in the single-celled organisms would come to power animals millions of times their mass.

A similar theme is found in the evolution of sensory systems. Many prokaryotic microbes evolved chemosensory systems that are analogues of taste and smell in animals. Vision is but one classic example of an apparent difficulty in transferring microbe-derived systems to more complex organisms. For many years, the evolution of eyes was viewed as so complex that they could have been formed only if guided by a divine creator. Indeed, Darwin appeared to puzzle over the evolution of eyes, but his musings on this subject were stymied by a lack of information. In the first edition of *The Origin of Species*, Darwin wrote,

> To suppose that the eye with all its inimitable contrivances for adjusting the focus to different distances, for admitting different amounts of light, and for the correction of spherical and chromatic aberration, could have been

formed by natural selection, seems, I freely confess, absurd in the highest degree. Yet reason tells me, that if numerous gradations from a perfect and complex eye to one very imperfect eye and simple, each grade being useful to its possessor, can be shown to exist, as is certain the case; if further, the eye does vary ever so slightly, and the variations be inherited, as is likewise certainly the case; and if any variation or modification in the organ be ever useful to an animal under changing conditions of life, then the difficulty of believing that a perfect and complex eye could be formed by natural selection, although insuperable by our imagination, can hardly be considered real.

Darwin had no idea that microbes evolved several sensors for light. In animal eyes, a pigment, *retinal* (which is derived from vitamin A), is bound to a protein, *opsin*. The opsins are a very large family of proteins, all of which share the same basic structure, seven helices, that span a cell membrane. In animals, the retinal-containing protein is the light sensor, but very similar pigments are bound to other opsin proteins in many microbes. Microbial *rhodopsins* are extremely common throughout the world's ocean. Were these two pigment protein complexes derived from a single common ancestor? The answer appears to be no. Opsins appear to have evolved independently at least two different times. In prokaryotes and some single-celled eukaryotic organisms, they often serve to pump protons, which are used to generate an electrical gradient across a cell membrane. These pigment protein complexes also have seven transmembrane helices, but their amino acid sequences are completely different from those in the opsins of animal eyes. In microbes, the pigment-protein complex is used to make energy. The microbes use rhodopsins to move protons across their cell membranes. The protons flow out through the whirling coupling factor, allowing the cell to make ATP in the presence of light. But the same pigment-protein complexes can also act as light sensors. In many single-celled eukaryotes, rhodopsins allow the cell to swim toward light of specific colors. The pigment was essentially retained and recycled with different proteins of remarkably similar structures in a wide range of single-celled eukaryotes, and then later, in animals, where it was bound to a different protein.

Eyespots, which are found in several single-celled eukaryotic algae, are crude optical sensors that contain rhodopsins. The genes for these opsins appear to have been horizontally transferred through several microbial lineages. Opsins are also found in corals, where the pigment-protein complex senses light, and then the animal uses that cue for spawning. In the evolution of true eyes, which not only sense light but also focus an image, a similar type of rhodopsin was layered across membranes. A lens made of collagen formed, and the optical "camera" eye was coupled to sensory systems leading to a complex organ, the brain, which could record images and compare them with previous records. In the embryological development of vertebrates, eyes are formed as direct extensions of the brain.

As discussed earlier, all living cells maintain an electrical gradient across their membranes. While the electrical gradient is critical to transporting nutrients from the environment into a cell and waste products out of a cell back to the environment, it also acts as a sensory system, allowing cells to sense gradients in light, temperature, or nutrients. In animals, special cells, *neurons*, evolved to coordinate behavior via transmission of electrical energy. In the evolution of animals, sensory systems—such as taste, smell, and vision—also generate an electrical signal and had to be coordinated with movement, so that an animal could catch prey, mate with the appropriate sex of their own species, flee from a predator, and learn.

These basic functions, which are critical to the survival of any animal, are derived from membranes in cells that evolved billions of years earlier. But to make the wires and brains within animals, significant innovations were needed. The cell had to gate the information—that is, turn on a switch for an electrical discharge and let the wire conduct a signal for only a moment. The signal had to have directionality—it should send the signal only one way along the wire, and not the other. And the cell had to communicate that signal to another cell to extend the wires or to coordinate the network—doing so required a chemical communication system. The chemical signals are based on simple molecules, many of which are derived from amino acids, and this communication system within animal cells builds on the foundation of quorum sensing in microbes. These evolutionary innovations led to neural networks, and finally to brains, which integrated information and controlled the wires in a two-way communication pattern—sensing and responding.

Neural networks and brain systems became increasingly complex as animal evolution continued. They are *emergent* properties. They are analogous to something along the lines of construction of the first computers, which were slow and had very little memory, but as computer scientists and engineers learned, they created faster, smaller, cheaper, and far more sophisticated systems. This basic process happened with neural systems in animals, and it led to massive changes in how Earth operates. But before exploring that, we have to understand the concept of symbiosis on a planetary scale.

The evolution of animals appears to have preceded the evolution of plants on land by approximately 200 million years; however, the two groups of organisms had very similar trajectories. Terrestrial plants are derived from a single group of green algae and began to colonize land about 450 million years ago. Deprived from a continuous source of water and nutrients, these early pioneers had to evolve a new set of characteristics to allow them to survive in the harsh, dry environment of land. Like animals, plants evolved a glue that allowed cells to stick to each other, but in their case, the glue is based on a polymer of sugars, *cellulose*, which is easy for plants to make. Cellulose does not require any nitrogen or phosphorus—only carbon, oxygen, and hydrogen, all of which are abundant. Additionally, cellulose and its derivatives are difficult for most microbes to break down. Animals cannot digest paper; only particular microbes in animals' guts can do that. Cellulose gives plants structural support on land, and when land plants die, some of the cellulose is incorporated into the soil and some is washed to the ocean, where it becomes incorporated into sediments.

Like the burial of single-celled photosynthetic eukaryotes 500 million years earlier, the evolution and death of land plants boosted the oxygen concentration in Earth's atmosphere—big time. Land plants were the biological Bolsheviks of their time. It is estimated that because of the rise and death of large land plants that were the forerunners of modern trees, the concentration of oxygen in Earth's atmosphere 350 million years ago was about 35%—or about 67% more than at present. What was the consequence?

The rise in atmospheric oxygen led to a massive invasion of land by animals from the sea. Worms, crustaceans, snails, and animals with backbones all successfully crawled up and colonized the new landscape.

Unlike the rise of plants, the emergence of animal life on land was the outcome of multiple invasions of many, many organisms. Except for the earliest animals to evolve—sponges, jellyfishes, and their relatives—almost every single animal body form has been successful in colonizing land.

The increased oxygen concentrations in the atmosphere, which was spurred by the success of terrestrial plants, allowed for several innovations in animals. The crustaceans and their relatives evolved to become insects. In insects, oxygen is supplied via diffusion through small openings along the sides of their bodies. Fossils of dragonflies with wing spans of half a meter are found from this period. Such large insects could not exist without extremely high oxygen concentrations. The earliest terrestrial fishes ultimately would evolve into amphibians and reptiles, and much later, into dinosaurs (including birds) and mammals. But that took a few more adjustments. Although marine animals had evolved systems to transport oxygen to their internal organs, allowing them to become bigger and more complex, the same circulatory system would not easily function on land due to massive water loss. Diffusion of oxygen in water is slow, but organisms could obtain the gas via direct exchange across their cells or via special organs, such as gills, which have extremely large surface areas. These gas-exchange systems could not easily work in air: the organism would quickly dry out. To help overcome this problem, the gas-exchange processes were internalized, and organisms evolved surfaces that prevented water from simply diffusing to the environment. Gas exchange was further accelerated by circulatory systems connected to a fluid that transported oxygen to distant parts of the organism. The circulatory systems required a pump to make the gas exchange process efficient—and rather than having a coordinated set of flagellated cells push fluid, as in a sponge, the molecular motors of single cells were co-opted for specialized cell functions, especially in muscles and neurons.

Muscles use enormous amounts of ATP to move the billions of myosin molecules across their tethers of actin every second. Neurons use huge amounts of energy to fire their cells. Compared with microbes, animals are the biological equivalent of jumbo jets in a world of recreational bicyclists. This may seem a paradox. If we were to take any animal and measure its energy consumption, it would be far lower than if the animal were plated across a giant Petri dish as individual cells one layer thick. That is because individual cells in animals are ultimately limited by the diffusion of oxygen. However, the total production of energy by animals is extremely high, even for cold-blooded animals such as turtles or snakes. For very active animals that have even higher body temperatures, such as birds and mammals, the energy demands are four to eight times greater than for reptiles.

All animals depend on photosynthetic organisms for their energy. In the oceans the food supply is overwhelmingly dominated by phytoplankton, which for most large animals are very difficult to harvest. The energy of phytoplankton is transferred to larger animals via smaller animals, such as small shrimp-like organisms, the zooplankton. That energy transfer comes with a cost. For each transfer up a food chain, only about 10% of the energy is retained in the next trophic level. For example, 100 pounds of phytoplankton would lead to the production of about 10 pounds of zooplankton, but the 10 pounds of zooplankton would only lead to the production of about 1 pound of fish. In the oceans, the concentration of phytoplankton is highest where nutrients from deep water are brought to the surface mostly by winds. These upwelling areas occur along continental margins and shallow seas—and that's why fisheries are so extensive in such areas. But the result is that the average life of a phytoplankton cell is five days. All the cells divide about once every five days, and one of the two daughter cells is eaten. The oceans contain only about 0.2% of the photosynthetic biomass on Earth. On land, however, most of the remaining 99.8% of the photosynthetic biomass isn't eaten. Most of the leaves on trees stay on the trees. But the same rule of trophic transfer applies on land as in the sea. One hundred pounds of grass will produce about 10 pounds of horse. However, because grasses tend to be fast growing and highly concentrated, bison can become large animals and form extensive herds. The number of trophic transfers in terrestrial ecosystems is generally smaller than in the oceans, and the evolution of grasses was a significant opportunity for the evolution of large mammals over the past 50 million years

A plentiful fuel supply led to huge competitive innovations in organisms' senses and the feedbacks to motors in the form of the evolution of sensors for smell, sight, taste, and sound. Animals evolved increasingly sophisticated systems to select plants to eat or prey to catch, and plants evolved increasingly sophisticated systems not only to use animals waste products to grow but also to use animals to pollinate their flowers and spread their seeds. The coevolution of plants with plants, plants with animals, and animals with animals led to an adaptive system of increasing complexity and more interactions.

To maintain a stable system of increasing complexity requires that each species adapt through time or its old evolutionary traits will become outmoded and the species will go extinct. Why? Because the environment is constantly changing on geological time scales, and natural selection is constantly at work.

The concept that organisms continuously evolve was playfully called the Red Queen hypothesis by an American evolutionary ecologist, Leigh van Valen, in 1973, after a story in *Alice Through the Looking Glass*. Van Valen's basic premise is that individual species have to "run in place" to maintain their evolutionary fitness. The oak trees we see today are not the same as the oak trees of five million years ago. This leads to an evolutionary game of hunt and fetch and to diversity via relatively small steps of biological innovation in a continuously changing ecological landscape.

Biological diversity in organisms is critical for ferrying the genes that code for the core, life-sustaining nanomachines across the vast landscapes of geological time fraught with existential hazards. But the diversity itself changes over time, and the evolution of specific traits has been adaptive only for short periods in the history of the planet. Organisms are transient vessels and are disposable. The genes are not.

One organism that evolved accidently but was selected because of very specific traits came to dominate the planet very rapidly in the recent past and has disrupted the planet like no other since the Great Oxidation Event 2.4 billion years ago or the evolution of land plants about 400 million years ago. In the landscape of large organisms with complex interactions, humans are the new animals on the planet and have rapidly become the new evolutionary Bolsheviks. We tend to think we are so different from other organisms that we can ignore the history of the planet. But can we?

CHAPTER 9

|||||||

The Fragile Species

In summers when I was a child, my father would often take me to Riverside Park, which was about a fifteen-minute walk from our apartment in the housing projects in Harlem. More than fifty years before my father was born in 1901, Riverside Park was a huge cemetery. It had been formally established as such in 1842 by an ordinance of the City of New York to accommodate the surge in deaths in the city from cholera, smallpox, and typhoid, which had led to the overcrowding of cemeteries further downtown. But although the ordinance allowed the city to later use Riverside Park as a massive cemetery for soldiers who died in the Civil War, the use of the area for burial of lost lives had a precedent more than a century earlier.

In an obscure little area across from Grant's Tomb, there is a small memorial to an "amiable child" who died in 1797 at the age of five. The gravesite is marked by a fenced, granite memorial to St. Claire Pollock, who is buried on a promontory overlooking the Hudson River and the cliffs of the Palisades in New Jersey. In 1797 that surely was a magnificent resting place, as the view would have been one of the most beautiful in the world.

As a child, I was very sickly and spent six months of my life in a hospital. I survived and have been well ever since, but I often thought about how that amiable child died and why children so young died so often so long ago. I also often thought about how fortunate I was not to have died in the hospital.

We humans have a long history of coexisting with microbes. Although some of our history is about peaceful coexistence, the peaceful aspects overlay a perpetual, low-level war between us and microbial invaders that are evolutionarily programmed to kill us. But we are not without a few of our own evolutionarily derived traits that give us some advantages in that war. In the course of human history, the war itself has greatly influenced our evolutionary trajectory and that of microbes. Let's consider one of the traits that give us some advantage in that conflict with microbes.

The evolution of complex language and abstract thought is one of the most interesting and important traits that distinguish us from all other animals, but it is only partially understood at a mechanistic level. A key evolutionary change appears to be two mutations between humans and our last primate ancestor that led to changes in two amino acids encoded by a *forkhead box* gene, *Foxp2*, which is found on chromosome 7 in our genome. The protein encoded by the *Foxp2* gene is a transcription factor that controls the expression of many genes during development of a fetus. In humans, this gene is critical for the development of several areas of the brain, including the Broca region, which is responsible for speech. Mutations in key areas of the *Foxp2* gene can lead to the inability to speak, articulate, or understand speech. This so-called language gene, which evolved from small and seemingly insignificant mutations between primates and humans, was transformative in our own evolution.

There are undoubtedly other genes that are involved in endowing humans with the ability to speak and to communicate complex, abstract thoughts to each other, but whatever they are, they allowed for a different mode of evolution, which anthropologists call *cultural evolution*; I prefer to call the phenomenon *horizontal information transfer*. The ability to communicate such thoughts rapidly is exceptional and exceptionally profound. Humans are the only animal that can transmit complex information across generational boundaries virtually instantly. Consequently, acquired knowledge can be preserved without any genetic selection. Horizontal information transfer potentially allowed humans to escape the Red Queen constraint. For example, if through horizontal information transfer we could control our exposure to, or life strategies of, microbes that can kill us, could we launch a counteroffensive and kill them first? In so doing, would we alter the evolutionary trajectory of microbes?

One could plausibly argue that humans and microbes have coevolved rapidly over the past 20,000 years, and perhaps even earlier. Certainly both we and microbes have benefited. For example, archeological evidence suggests early hunter-gatherer tribes had the capability of fermenting grains to make some alcoholic beverages, perhaps a beer. Naturally occurring microbial yeasts convert sugars in the grain to alcohol. By 3500 BCE, beer was a popular drink in Samaria and other areas in the cradle of civilization. Similarly, there also is evidence that wine predates written history. Archeological evidence suggests that it was produced in China by about 7000 BCE; by 3200 BCE wine was produced throughout the Middle East. Fermentation of grains and fruits to make alcohol eventually became widespread throughout Asia and Europe. It was the beginning of a boom for microbes in human culture.

Microbial fermentation processes were developed independently by many cultures and applied to many foods to produce cheeses, to modify soybeans (for example, to produced miso paste and soy sauce), and to create many other products from beans, cereals, fruits and vegetables, fish, and even meats.

The fermentation process is an example of our "peaceful" coexistence with microbes, and it has served at least three purposes from a human perspective. It allows much longer shelf lives of foods. That was especially important when food supplies were tied to seasonal availability and when other means of preservation were not readily accessible. Fermentation also often yields foods of higher nutritional value. Through human selection for taste or other attributes, specific microbes have been cultured in human foods long before it was understood that these organisms were responsible for the fermentation process. Fermentation also helps make foods more digestible. Microbes break down indigestible materials and make them accessible for human consumption. Cocoa and coffee beans are examples of foods in which the pulp surrounding the beans is naturally degraded by microbes before the beans are ingested and further processed in our guts.

Microbes are at the head of the class that can perform the tricks that humans would want to select for. A very small subset essentially became invisible "pets" for performing their unique tricks, for example, by converting one specific sugar to a specific acid to make a certain cheese or specific beer or bread, and so on. But sometimes these "good" microbes are out competed by other microbes, and food becomes toxic, making us sick and even killing us.

In centuries past, premature death from microbial infections was so common that it was assumed that more than half of the children born in any family would not survive to an age of reproductive capability. For example, during the sixth century, an outbreak of bubonic plague, which is caused by the bacterium *Yersenia pestis* and is transmitted by flea bites, killed approximately 50 million people in the Byzantine empire of Justinian I. In the fourteenth century, another plague pandemic led to the death of approximately 50% of the entire population of Europe. Outbreaks of bubonic plague continued well into the seventeenth century in England, Italy, and Spain.

In the nineteenth century, cholera pandemics, caused by infection with the bacterium *Vibrio cholera*, were extremely common throughout Asia and killed tens of millions of people. The disease, which is spread by fecal contamination of drinking water, pervaded Europe, killing many millions in Hungary, Russia, Britain, and France, and even came to the United States via immigration. Cholera killed James Polk in June 1849, three months after he left office of the president of the United States. Huge numbers of people were killed in the nineteenth century by typhus, smallpox, tuberculosis, pneumonia, and influenza. Clearly the threats from microbes to human health are enormous.

Microbes get into our bodies via our mouths from food and water, into our lungs via the air we breathe, from sex, animal bites, and even from cuts. They wreak havoc with our respiratory, circulatory, and digestive systems and cause major infections that are easily transmitted across wide swaths of human populations. Microbes can produce extremely potent neurotoxins, enterotoxins, and myriad other molecules that target specific functions. Sometimes we can control the toxic effect, such as when we use botulinum toxin, which targets the neurons and muscles, as a therapeutic and cosmetic medication to reduce muscular spasms and wrinkles. However, more often the actions of these very potent toxins are difficult to control once the microbes enter our bodies. In short, until the twentieth century, by killing large numbers of us, microbes largely kept the populations of humans under control. Although microbial infections still affect many people, particularly in the underdeveloped and developing worlds, two major breakthroughs changed the relationship we have with microbes.

The first was the recognition that by minimizing exposure to specific microbes, disease could be avoided. To this end, one of the greatest changes in the exposure to microbial toxins was in how water is delivered to and removed from homes. The threat of waterborne diseases was greatly reduced over the centuries both by treating water and reducing human exposure to sewage. Boiling water with herbs or other flavoring agents became common throughout Asia, as did the addition of alcohol derived from the fermentation of grains and fruits. These two processes were used for centuries in various incarnations to make water safe to drink. Sewage disposal systems came much later, and they further greatly reduced the risk of exposure to microbial diseases. The knowledge base about water supply and waste removal quickly spread in the nineteenth century and is a hallmark of developed countries.

The second breakthrough was the discovery of natural metabolites that kill microbes. *Antibiotic* is a term that was coined by the late Selman Waksman, who discovered streptomycin, a molecule which is produced by a microbe that was isolated from a small sample of soil found right outside my laboratory. That discovery allowed countless millions of sick people to become well. It is virtually impossible to find an adult in a developed country who has not had a course of antibiotics in their lifetime.

In the mid-twentieth century, it was also discovered that giving antibiotics to animals led to increased meat and milk production. Approximately 80% of all the antibiotics consumed in the United States are used for animal production, not human health. Indeed, so many antibiotics are currently applied, especially in animal husbandry, that many microbes have become immune to common antibiotics—and are fighting back to kill us. Their immunity is due to mutations. Because microbes can replicate very quickly, on the order of hours or less, natural mutations accumulate rapidly; the mutations are then selected by our application of antibiotics. The microbes that live are survivors, and once selected, they rapidly spread across myriad microbial communities via horizontal gene transfer. These virulent microbes have launched a counteroffensive against us. In effect, the microbes are fighting back in what is turning out to be a Red Queen evolutionary cycle of escalating defense on our part leading to escalating offense on the part of microbes.

Regardless of who the ultimate winner in the Red Queen cycle is, human knowledge, which is acquired and disseminated globally by horizontal information transfer, has clearly been extremely effective in helping humans temporarily control the planet. Our ongoing war with microbes has led to great victories for humans. Although microbes have become increasingly resistant to antibiotics, their constraints on human life, while not insignificant, are far less influential than only a century ago. The evolution of language and the rapid transfer of information helped reduce microbial control of human population growth. We appear to have temporarily escaped the Red Queen constraint and in so doing have entered an exponential growth phase of human population.

As an undergraduate student, I worked in a microbiology laboratory at City College of New York, growing algal strains for experiments. In the laboratory, the growth of a single microbe in a culture with a broth of nutrients follows a simple trajectory. For some period after inoculation the cells grow slowly; this is called the lag phase. But after a bit, the cells start to get used to their new environment and grow faster. During this phase, the trajectory of the growth of the population is exponential: two cells become four, four become eight, and so on. Eventually, some nutrient in the media becomes limiting, and cells start competing with each other for the limiting resource. When that happens, the growth rates slow down and the population reaches a plateau.

There is also a fourth stage, which is seldom discussed in texts. When cells have reached their plateau and are nutrient limited for a period of time, they can have a difficult time making the basic nanomachinery for survival. Many of them "commit suicide." That phenomenon, which I accidently discovered many years ago as a graduate student but did not think about for many more years, is called autocatalyzed cell death.

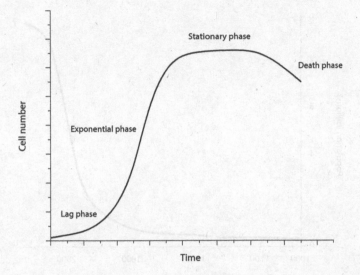

FIGURE 33. A typical growth curve for microbes. Upon inoculation, the cells undergo a lag phase before beginning to grow exponentially. At some point, a nutrient or other resource (for example, light in the case of algae) becomes limiting and the growth rate declines, eventually stopping. This is the stationary phase. If left for a long period of time without replenishment of nutrients and dilution, the cells will start to die.

This basic growth trajectory is more complicated in the real world, where many, many other microbes are inevitably competing for the same resources and predators and viruses are always present to keep any individual population of microbes in check. In the real world, individual species very seldom escape an exponential growth phase to dominate the ocean or landscape, unless they are an introduced species with no predators or else have other unique features that allow them to outcompete the indigenous organisms.

The basic concept of checks and balances in growth of microbes applies to any organism, including us. In year 1 CE of the Gregorian calendar, it is estimated there were between 250 and 300 million humans across the globe. In 1809, when Darwin was born, there were about 1 billion people on Earth. By the end of the nineteenth century, there were about

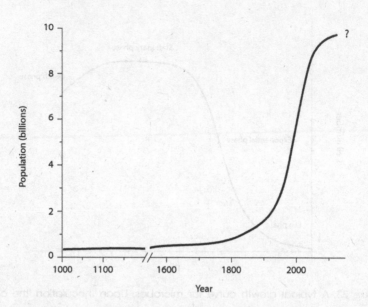

FIGURE 34. A growth curve of the human population since the year 1000 CE. Prior to the Industrial Revolution and the discovery of how to separate sewage from clean drinking water, the human population was relatively constant, analogous to the lag phase in a microbial culture (Fig. 33). From the middle of the nineteenth century, however, the human population has grown exponentially. Demographers estimate that it will plateau in the middle of the twenty-first century at approximately 9.5 to 10 billion people. Compare with Figure 33.

1.6 billion people, and the global average life expectancy was only about 30 years. By the end of the twentieth century, there were more than 6 billion people, and global life expectancy had more than doubled, to about 65. By 2050, it is estimated that more than 9.5 billion humans will inhabit this planet, each one of which will require food, water, energy, and fibers. Demographers hope that that is the plateau of human population, but no one can be absolutely certain.

Given massive population growth, how can we sustain ourselves? Something is ultimately going to limit our population. Will it be food? Water? Energy? Space? Will microbes become increasingly resistant to our most advanced antibiotics and once again be able to kill us en masse?

Or will we permanently distort the microbe-controlled chemistry of our planet so that it is less hospitable for humans?

Let's consider a small incident that led to a massive alteration of our planet—by us.

In 1859, the same year that Big Ben chimed for the first time and the London publisher John Murray and Sons sent the first edition of *The Origin of Species* to press, on the other side of the Atlantic an American train conductor, Edwin Drake, drilled the first major oil well near Titusville, Pennsylvania. That event would come to mark the beginning of the modern boom in oil exploration and, ultimately, exploitation. At the time, petroleum (literally, "rock oil") had a limited market. Its primary use was to make lamp oil, kerosene.

The kerosene lamp had been developed in the United States by Robert Dietz, a small-time inventor in Brooklyn who owned a factory that made oil lamps. Dietz had designed a lamp that burned brightly with very little smoke. His lamps were as transformative in their time as the invention of the incandescent light bulb forty years later, but when they were first developed, Dietz lacked a source of cheap fuel. The primary source of lamp oil at the time was derived from whale blubber, particularly from sperm whales. The Titusville well provided a new source of oil from which to make kerosene. Coupled with Dietz's marketing of kerosene lamps, there was a dramatic expansion of the lamps across the country. The rise of the new technology of the time led to a reduction in the demand for whale blubber, with the unintended consequence of the collapse of the whaling industry in the latter half of the nineteenth century. While the use of kerosene as a fuel for lighting can be claimed as saving whales from being hunted to extinction, there were other unintended consequences to follow.

By the early decades of the twentieth century, the oil industry had become the engine of economic growth in the rapidly industrializing nations. One by-product of distilling kerosene was a very volatile liquid, gasoline, which had no market at the time, so it was burned off as a waste product. However, by the end of the nineteenth century, several people had developed, in one form or another, the internal combustion engine. In 1876, after more than a decade of experimenting, a German engineer, Nikolaus Otto, with the help of numerous colleagues, succeeded in developing an internal combustion engine that was capable of running on petroleum distillates. Gasoline was so cheap that it quickly became the obvious fuel to use. Gasoline-powered engines were far more efficient than coal-fired steam engines or coal-gas engines and were quickly adopted for use in transportation. The new engines led to a huge demand for the waste product of the kerosene industry, and to meet that demand, oil companies invested massively in infrastructure to refine petroleum and transport fuels.

However, an unintended and totally unforeseen consequence of the rapid combustion of petroleum and other fossil fuels was the rise in greenhouse gases, especially carbon dioxide. For every gallon of gasoline burned, approximately 20 pounds of carbon dioxide are emitted from the tailpipe of a car. There are more than one billion cars on the road worldwide, and that is only part of the problem. There are vast supplies of coal and natural gas across the globe. All of these fossil fuels were produced millions of years ago and represent a reservoir of stored energy bonds. In the case of petroleum in particular, they are stored bonds made from the remains of fossilized algae. We have developed very efficient systems for extracting the fuels. In one year, we can extract one million years' worth of petroleum—or, to put it another way, it took one million years of photosynthesis of algae and higher plants to make the fuels we burn in a single year.

Since the beginning of the Industrial Revolution, in the mid-nineteenth century, atmospheric carbon dioxide concentrations rose exponentially from 280 parts per million in 1800 to more than 400 parts per million in 2014, and there is no plateau in the foreseeable future. The continued reliance on fossil fuels greatly increases the potential for long-term global climate change, including warming and acidification of the upper ocean, loss of glacial ice, rise in sea level, and increased frequency and intensity of storms. We have begun to produce a waste product of our own making that severely impacts the planet, but we don't know an easy way to fix the problem. Can we develop renewable, carbon-neutral fuels that are environmentally sustainable, are economically viable, and can directly displace petroleum based products using the existing infrastructure? As

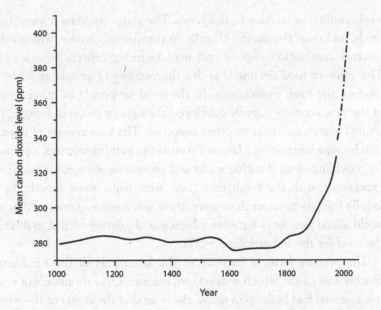

FIGURE 35. The change in concentration of carbon dioxide in Earth's atmosphere since 1000 CE. Until the Industrial Revolution, atmospheric carbon dioxide concentrations were relatively constant at approximately 280 parts per million by volume (i.e., 0.028%, compared with oxygen at 210,000 parts per million, or 21%). Since the Industrial Revolution, the concentration of the gas has risen almost exponentially, and in 2014 reached 400 parts per million. Unlike nitrogen and oxygen, carbon dioxide is a greenhouse gas and traps heat. The relatively small concentration of this gas in Earth's atmosphere is critical to controlling climate. The curve for the change in carbon dioxide is strikingly similar to the growth curve for the human population (Fig. 34).

we will shortly see, we have placed a lot of hope on microbes to help save us. But even more unintended consequences were to follow the fossil fuel problem.

The development of fossil fuels led to a massive change in how we grow, harvest, process, and transport our food. Fields, which previously had been plowed by oxen or horses, could now be plowed by machines driven by internal combustion engines fueled by petroleum. The harvesting of wheat, corn and other bulk crops, which once required back-breaking work, could now be done by machines. The grains could be shipped hundreds, and even thousands, of miles to population centers, where other internal combustion engines and manufacturing centers were located. The price of food declined, as did the number of people required to produce the food. Simultaneously, the need for people in other sectors of the new economy rapidly developed throughout much of Europe, the United States, and later in other countries. The new centers of population became large cities. Massive investments in infrastructure, especially for providing clean drinking water and processing sewage, increased life expectancy, with the result that there were many more mouths to be fed. By the late nineteenth century, there was a serious concern that the world would run out of fertilizer, which was absolutely critical to produce the food for the industrialized world.

The primary form of fertilizer in the latter part of the nineteenth century was guano, which is dried bird manure. Over thousands of years, this material had built up in many places around the coasts of the world, and the export of guano from Chile, Florida, and several other coastal regions was a major industry. But with a growing human population, guano was being used faster than it could be produced by birds. The cost of guano began to increase, and it was realized that a replacement was needed. But with what could it be replaced?

One of the most important plant nutrients in guano is ammonium and other so-called fixed nitrogen products. The nitrogen was originally "fixed" into compounds by microbes in the oceans and found its way into algae, then into small animals, and ultimately into fish, which the birds ate. In the latter part of the nineteenth century, we didn't really understand what nitrogen fixation was. It was not until 1901 that a Dutch microbiologist, Martinus Beijerinck, showed that bacteria associated with the roots of legumes could convert gaseous nitrogen in the air to a form that a plant could use to grow. While crop rotation could help restore nitrogen to soils (and that technique is still used), it was realized that without the addition of external fixed nitrogen, we could not grow enough food to feed ourselves.

In 1898, the newly elected president of the Royal Society in London, the same august body that had published Robert Hooke's *Micrographia*

274 years earlier, laid out a challenge: "Find a replacement for ammonium" to save "England and all civilized nations." Sir William Crookes, a well-known Victorian scientist who had discovered a new element, thallium, (and was a spiritualist), was concerned that unless humans could fix nitrogen for agriculture, the civilized world would starve by the 1930s. By "civilized" Crookes meant people who ate wheat rather than "inferior" grains, such as rice. It was not clear how organisms fixed nitrogen, but it was clear that guano supplies wouldn't last forever. Crookes' challenge was taken up by chemists.

In Germany, a cantankerous German Jewish chemist, Fritz Haber, worked patiently to find a chemical catalyst that could take the relatively inert gas, nitrogen, which forms 78% of Earth's atmosphere, and combine it with hydrogen under high temperature and pressure to produce ammonia, which when dissolved in water becomes an ion, ammonium. After several years, Haber succeeded in producing about a glass of ammonia per hour with a machine about the size of a large box. It certainly didn't look like a big deal, but the reaction worked. The catalyst was based on iron and was not difficult to synthesize, but to bring the reaction onto the market required a major investment. Haber wasn't interested in marketing anything, let alone ammonia; he was a scientist.

To Carl Bosch, a chemical engineer working for the German industrial chemical company BASF, Haber's invention was an inspiration. He convinced the upper management of BASF to develop a pilot plant, which required a lot of energy to produce ammonia, but it nevertheless worked. The hydrogen gas for the reaction was derived from coal, which also was used to heat the two gases in the reaction vessel to make ammonia. Germany had lots of coal, and BASF was on its way to becoming very rich from owning the secret pathway for producing fertilizer. The Haber-Bosch reaction remains, with minor modifications, the backbone of the world's supply of fixed nitrogen for fertilizer to this day. Without that process, we almost certainly could not feed 7.5 billion people or even think about feeding another 2 billion by the middle of the twenty-first century.

In effect, humans have developed massive machines to fix nitrogen, bypassing the need for nurturing the nanomachines nature designed for exactly the same process in microbes billions of years earlier. Our human-fabricated machines—the planes, trains, and cars, the nitrogen-fixing factories, sewage treatment facilities, the steel mills, and all the other energy and material intensive processes—are relatively recent creations. Virtually all were designed during the past two centuries, since the beginning of the Industrial Revolution, but they were not designed to be compatible with the biogeochemical processes established during the past several hundred million years of Earth's history. The result of these human-made machines has been rapid alterations in the chemistry of the planet. It will take several hundreds, if not thousands, of years for microbes to restore Earth to a new equilibrium.

Nitrogen fixation by humans greatly exceeds that of all microbes on the planet, and fixed nitrogen pours off fields all across the world into rivers and out into the coastal oceans, where it stimulates algal blooms. The algal blooms often are so large that when the organisms sink, die, and are consumed by other microbes, massive loses of oxygen ensue, fish die, and gases such as nitrous oxide, laughing gas, are emitted.

Laughing gas is no laughing matter. Each molecule of nitrous oxide has 300 times the heat-trapping capacity of carbon dioxide; it is a very potent greenhouse gas. However, there is another part of the problem related to maintaining Earth's balanced market for electrons on a planetary scale.

In World War I, as Germany was fighting the French and British, gunpowder started to become scarce. A key ingredient of gunpowder is saltpeter, which is a potassium salt of nitrate. Nitrate is another fixed-nitrogen molecule that is formed when microbes combine an ammonium ion with three oxygen atoms. There are very few places in the world where nitrate can be mined. The salts of nitrate are very soluble in water, and when it rains, nitrate becomes dissolved in the rainwater and flows into soils or runs into rivers and lakes. A major source of nitrate for Germany was a natural reservoir in the Atacampa Desert in Chile, the driest desert in the world.

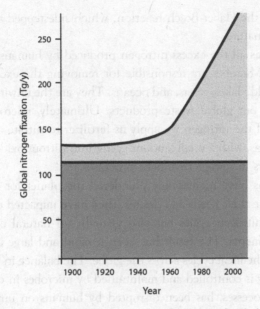

FIGURE 36. The change in the total amount of nitrogen fixed during the past century. Prior to the invention of the Haber-Bosch reaction for fixing nitrogen, all nitrogen was fixed by microbes with a small contribution from lightning. Natural biological nitrogen fixation is approximately 100 teragrams (10^{12} grams) per year (darker area). After the introduction of the Haber-Bosch reaction, human production of fixed nitrogen increased dramatically and presently exceeds natural biological nitrogen fixation by almost a factor of two (lighter area).

Germany had to protect its supply of nitrate as it was transported from South America to Europe. In 1915, during the First World War, the British navy destroyed the German naval vessels that protected the nitrate supply. Germany's supply of nitrate was stopped, thus halting the production of gunpowder and creating a shortage of ammunition. It was potentially a key factor in Germany's defeat in the First World War. Regardless, when Hitler took power of Germany, he demanded that BASF find a pathway to convert ammonia to nitrate. German chemists obliged, and the primary source of fertilizer on the world market to this day is ammonium nitrate, which does not exist in nature (and is extremely explosive). The production of ammonium nitrate was an extension of the Haber-Bosch reaction, which sidestepped all microbial reactions in nature.

Where does all the excess nitrogen produced by humans for food go in the end? Microbes are responsible for removing the excess nitrogen from the world's lakes, rivers, and oceans. They are the unwitting garbage recyclers for our global waste products. Ultimately, microbes convert about 25% of the nitrogen we apply as fertilizer to nitrate and then on to nitrogen gas, with a small amount going into nitrous oxide. The same process occurs in sewage treatment processes.

As humans have increasingly plundered the planet for resources to feed and serve their needs and desires, they have impacted not only the carbon and nitrogen cycles but also virtually all natural cycles of the chemical elements. The result has been a rapid and large distortion of basic biogeochemical cycles across the globe. The balance in these cycles, which largely is controlled and maintained by microbes in concert with geological processes, has been disrupted by humans on unprecedented scales over a very short period of time. The result is that the natural cycles of carbon, nitrogen, sulfur, and many other elements are *decoupled*, by which I mean that the changes in the cycles are becoming increasingly independent of each other. For example, prior to human evolution, the carbon and nitrogen cycles were intimately interconnected. There was no massive flow of nitrogen into rivers. In the industrial world, the production of ammonium is not strictly related to the rate of combustion of fossil fuels.

Is there an "off ramp"? Can humans cohabit the planet with microbes without plundering so many resources and disrupting its chemistry so rapidly? If so, how can we get on that path?

One approach, taken increasingly seriously, is to engineer microbes to do our bidding. A field of science has emerged, *synthetic biology*, in which scientists try to design the metabolism of microbes so that they can fix nitrogen orders of magnitude faster than they would naturally, or try to produce a replacement for petroleum, or try to engineer a protein that can be a feedstock for an artificial meat. The limits appear only to be in our imaginations. Let's see how this type of approach has elicited hope.

CHAPTER 10

||||||||

The Tinkerers

Increasingly over the course of human evolution, we have become control freaks. For thousands of years humans have bred and selected animals and plants, cleared land, created new materials, and built structures. We have diverted rivers to control the flow of water across continents and built walls to hold back the sea. We have engineered machines to transport food, materials, and ourselves across the planet. It should not be surprising then that over the course of a few short decades, we have also come to engineer microbes. As we will see, scientists want to transfer, enhance or silence genes to make microbes work for us without having to hassle with natural selection. We will be the creators of microbial metabolism and will design microbes to do our bidding. We have the power to do so, but that power does not appear to come with an understanding of the potential tremendous consequences for microbial evolution, let alone for our role in altering the future trajectory of the planet.

For more than two decades I worked at a U.S. government national laboratory funded largely by the Department of Energy and its legacy agencies. National laboratories were conceived of and designed to bring to fruition big ideas in physics and chemistry, and many people correctly associate them with the invention and production of atomic weapons, which was their initial intent. However, the national laboratories also often have huge computers and other machines, such as high-energy colliders, designed to understand the nature of matter, extraordinarily powerful microscopes, and engineers who work with scientists to develop technologies that lead to new discoveries.

I went to lunch every week with chemists and physicists who had worked on the atomic bomb with Oppenheimer, Fermi, Urey, and Seaborg. For the most part, most of my lunchtime colleagues viewed biology as something of an afterthought. Unlike physicists, biologists seldom needed machines that cost tens, if not hundreds, of millions of dollars to build. They didn't think on grand scales, like the physicists or even the chemists. But in the early 1980s, a grand challenge in biology was proposed by a few scientists at the Department of Energy: sequencing the human genome. The basic idea was to develop technologies that could rapidly and cheaply sequence the genomes of organisms and to turn the sequences into useful information.

The initial response was not very positive. The idea was not based on a specific hypothesis, which is how most biologists frame their research, but rather on the desire to collect and analyze lots of genetic data. But the idea, when it finally caught on, not only transformed our understanding of human genomes, it also transformed our understanding of microbes in the environment. It rapidly altered the nascent field of *molecular biology* and made it one of the cornerstones of biological research ever since.

There were many scientists who contributed to the development of the field of molecular biology since its early, exponential growth phase, and inevitably a recounting of the history would be fraught with omissions. However, three main discoveries, aided greatly by other fundamental discoveries in the twentieth century, facilitated our ability to deliberately transfer genes horizontally in microbes and thereby potentially alter the course of evolution. The concept of horizontal gene transfer is simple—as we saw earlier, microbes move genes from one organism to another all the time. But the concept that humans could move genes from one organism to another without the messy problems of sex and natural selection meant that we could potentially "design" microbes. My selection of the key events that led to the development and maturation of genetic engineering is based on history as it reflects our future as a species and our subsequent investment in microbes to be our saviors.

One of the most important discoveries was made by Oswald Avery, a Canadian-born physician who, at the Rockefeller Hospital (now a part of the Rockefeller University—the same place Palade, the discoverer of ribosomes worked) with Colin MacLeod and Maclyn McCarty, reported in 1944 that DNA was the carrier of genetic information. The early experiments were rather simple but quite profound. Avery and his colleagues used the technique of *transformation*, which had been discovered in 1928 and is the cornerstone of horizontal gene transfer experiments to this day. Transformation was discussed earlier in the context of horizontal gene transfer in consortia but not the details of how it works.

For many years, microbiologists understood that there were several strains, or *serotypes*, of microbes with a common genetic background. Indeed, in the case of *Escherichia coli*, which was originally discovered in 1895 by a German physician, Theodor Escherich, in healthy human feces, it was later discovered that some variants of apparently the same bacterium led to death if ingested. Similarly, a British microbiologist, Frederick Griffith, realized that a bacterium responsible for pneumonia in humans, *Streptococcus pneumoniae*, was present in healthy adults and did not cause the disease.

Griffith had isolated a virulent strain, killed the microbes with heat, and injected them into mice. The mice lived. But if he mixed the heat-killed virulent strain with a nonvirulent, living one and injected the microbes into mice, they died. Griffith had no idea what was going on at a molecular level and called the phenomenon a "transformation phenomenon." In essence, Griffith could transform a nonvirulent form of a microbe into a virulent form with a suspension of *dead* virulent microbes. It was almost like magic. He published his findings in 1928 and listed his affiliation as "from the Ministry's Pathological Laboratory"—clearly the irony of the word "pathological" has also evolved in the past century.

Oswald Avery was extremely skeptical of Griffith's experiments and set out to repeat them. After a long period he concluded that Griffith, who was a meticulous researcher, was right. So what was happening?

To deduce the identity of the transformation agent, Avery and his colleagues incubated the broth containing the dead bacteria isolated from a virulent strain with enzymes that digested proteins. At the time, most biochemists thought that proteins were the carriers of genetic information, because they were found in chromosomes of eukaryotic cells and, being composed of twenty different amino acids, comprised sufficient variability to account for genetic traits; thus, it was logical that these molecules carried the key to genetic information. Avery and his coworkers repeated Griffith's experiment, but with a twist: when they incubated the heat-killed virulent strain of the microbes with enzymes that digested proteins or RNA and then injected the solution into mice, the mice died. But when they added an enzyme that digested DNA, the mice lived. He concluded that DNA carried the genetic information from the dead, virulent strain to the nonvirulent strain. It was a remarkable discovery because it started to focus the scientific world on the nature of DNA. Equally remarkable, at that time, and in his day, Avery and coworkers were largely unappreciated, to say the least. Their work was almost ignored. The predisposition that proteins were the carriers of genetic information was so great that the publication by Avery and colleagues was looked on as an experimental artifact. It is an example of cognitive dissonance in modern academia. Many biochemists believed that the transformants Avery and colleagues had produced were contaminated by trace amounts of proteins.

Enter Joshua Lederberg, the genius son of a rabbi, born in New Jersey, who grew up in Washington Heights in New York City and spent most of his youth in libraries. He took Avery's papers seriously and set out to find the transformation factor, and in so doing, Lederberg literally transformed biology by revealing the "magic" of microbial transformation. He and his wife, Esther, used viral particles to insert genetic information into a bacterium—a process we now call *transduction*, which became one of the hallmarks of genetic engineering. The process is based on insertion of a circular piece of DNA into a bacterium—what Lederberg called a *plasmid*. A plasmid would self-replicate inside a bacterium, but only outside of the latter's chromosome. It was a foreign invader that could co-opt the replication system of the bacterium to replicate the foreign molecule inside the host microbe. Lederberg found that plasmids could allow the host bacterium to resist death by antibiotics. With that discovery, Lederberg became the pioneer of human-designed horizontal gene transfer in the laboratory—and that gave humans a new method for disrupting microbial evolution. Lederberg went on to win a Nobel Prize at the age of thirty-three.

On the basis of Lederberg's and other's contributions, biologists could now conceivably insert genes at will into virtually any organism. In principle, humans could become masters of the biological universe. Organisms' genomes were to become hunted like prey for our benefit—so we could find drugs or genes that made us live longer by resisting or curing diseases. (Somewhat ironically, Lederberg died at the age of eighty-two of pneumonia derived from a microbe he first studied as a student.) But in order to design organisms by transforming them, we needed to understand how the DNA codes for specific proteins. As *gene* designers, we needed to understand how nature made genes.

The discovery of the structure of DNA is legend, and legendary. DNA is a polymer of only four repeating cyclic molecules, nucleotides, linked by a five-carbon sugar through phosphate bonds to form a chain. The only variations in the chain were in the bases—and given that there were only four, DNA seemed boring. But if Avery and Lederberg were right, then the structure of DNA should reveal the "magic." But it didn't, at first.

The fundamental structure of the molecule was based on a single X-ray diffraction photo taken by Rosalind Franklin and Raymond Gosling at King's College, London, in 1952. On April 25, 1953, the venerable British journal, *Nature*, published a series of three back-to-back papers. The first, written by Francis Crick and James Watson at Cambridge University, proposed a structure of DNA based on the yet unpublished X-ray images that were made by Wilkins and Franklin. The second, independent, paper was from the laboratory of Maurice Wilkins at King's College in London and had a crude X-ray image of the molecule. The third paper, by Franklin and Gosling, showed a higher-resolution diffraction pattern, which they had obtained themselves. All three papers concluded that the molecule was probably a helix, but Watson, Crick, and Wilkins proposed that it was a double helix. The Noble prize for the discovery of the structure was shared by Crick, Watson, and Wilkins in 1962. Franklin had died of ovarian cancer in 1958, at age 37, and so was not eligible to receive it.

At the time, it was clear that the DNA molecule was the key to the inheritance of information. Somehow it encoded the sequence of the amino acids in proteins, but it was not at all obvious how the structure of DNA, as reconstructed from the analysis of the X-ray diffraction of the molecule, could contain the information for synthesis of proteins. There are only four different nucleotides in DNA. How could four nucleotides encode a system of information that led to the formation of proteins with twenty amino acids in very specific sequences?

The elucidation of the genetic code was, perhaps, even more ingenious than the elucidation of the structure of DNA. Following the work of Avery and his colleagues and the structural analysis of the double helix by Franklin, Gosling, Wilkins, Watson, and Crick, it was quickly realized that with only four nucleotides in DNA and twenty amino acids in proteins, there had to be more than one nucleotide that coded for an amino acid. The smallest number had to be three nucleotides. This logic was based simply on math. All the possible combinations of only two nucleotides yields $4^2 = 16$ amino acids, which are less than sufficient. If however, there are three nucleotides, the possible combinations are $4^3 = 64$, which are more than sufficient. Using a technique of inserting and deleting a single nucleotide in a virus that infected E. *coli*, a team led by Francis Crick and that included an iconoclastic scientist, Sydney Brenner, worked out the genetic code in the bacterium. They showed that a set of three nucleotides in a very specific sequence of DNA specified a particular amino acid. Their work literally translated the code, the Rosetta stone, for the understanding the inheritance of life. However, there were complications.

For most amino acids, more than one set of three nucleotides in a sequence encoded the same amino acid. From knowledge of the DNA sequence, one could deduce the amino acid sequence of the protein that the gene encoded. But the information was *degenerate*—that is, one could not deduce the exact DNA sequence from knowledge of the protein sequence. Knowing the "words" in one language in the DNA world specified one meaning in the amino acid world of proteins. But knowing the "words" in the amino acids of proteins did not allow a faithful translation to the DNA. The big problem for understanding how any living organism worked appeared to be lie in what instructions are encoded in the DNA. That problem led to another technical challenge—how could DNA be sequenced?

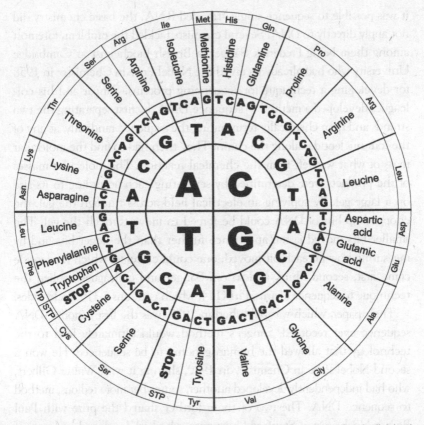

FIGURE 37. The codon wheel. This is the Rosetta stone of how the individual bases, or nucleotides, in DNA encode for specific amino acids in a protein. The code for each amino acid is contained in a sequence of three nucleotides called a codon. Starting in the center of the wheel and working out, one can determine which amino acid is encoded from the sequence of DNA. For example, a sequence of AGC encodes for the amino acid serine, whereas ACC encodes for threonine. With the exception of methionine and tryptophan, all amino acids have more than one possible codon.

Proteins, RNA, and DNA are polymers, and sequencing any biological polymer poses a huge challenge. The reaction has to cut off each monomer of the parent polymer in a specific order. Sequencing DNA was originally even more difficult because the polymer is double stranded, and although it was possible to sequence single-stranded RNA, the basic chemistry did not apply directly to DNA. Several chemists tackled the problem, foremost among them being Frederick Sanger, a British biochemist at Cambridge University who had already received the Nobel Prize in Chemistry in 1958 for developing a technique for sequencing proteins. Sanger and his colleagues developed a method to sequence DNA, by first separating the two strands and then chemically terminating the sequence randomly, at any of the four nucleotides along the chain. They then had to find the molecular mass of what was left from the chemical reaction. The molecular masses of the products were determined by separating each according to its size on a large gel. By applying an electrical field across through the gel, the chopped up bits of DNA could be forced to move through the gel. The smaller bits moved faster, and hence further than the larger bits and by measuring how far each bit moved, one could calculate which nucleotide came first, second, third, and so on. Sanger and colleagues applied the technique to sequence a virus, PhiX174, which contains 5,375 nucleotides.

Their paper, which was published in 1977, was the first genomic DNA sequence ever recorded. Sanger's method would ultimately lead to the technology that allowed the human genome to be sequenced. He won a second Nobel Prize in Chemistry in 1980, sharing it with Walter Gilbert, who had independently developed another, somewhat more tedious, method to sequence DNA. The two of them further shared the prize with Paul Berg, a biochemist at Stanford University, who had developed a process of making DNA molecules derived from two or more sources—molecules that do not exist in nature. Those human-constructed DNA molecules are called *recombinant* DNA. These three men helped to change the world perhaps as much as, or possibly more than, the discovery of the structure of DNA.

The basic chain-terminating sequencing methodology developed by Sanger cannot be applied to large sequences of DNA. To approach the problem of sequencing a human genome, which contains 23 chromosomes, the DNA had to be cut into smaller chunks. The individual pieces could be sequenced, and random overlaps between the sequences then matched to reconstruct the entire genome. This technique, which was given the designation *shotgun* sequencing (a term coined by Sanger), was developed earlier for microbes and applied to the human genome by J.

Craig Venter and colleagues. Indeed, while the technical aspects of the sequencing were difficult enough, reconstructing the order of the genes on each chromosome was even more challenging. The effort, which took several years to complete, revealed that our genome contains more than 3.2 billion base pairs, but only about 1.5% of them encode for proteins. This was one of the biggest surprises of the human genome sequencing project—we only have about 20,000 protein-coding genes, far less than predicted before the genome was sequenced and only a factor of two higher than simple worms. Thus, more than 97% of our genome contains noncoding regions, which are not present in microbes. Perhaps ironically, sequencing the human genome revealed how relatively small genetic changes can lead to the higher organizational patterns of an animal. The basic instructions for fabricating the machinery that supplies us with energy and allows for protein synthesis, transport of ions, and basic metabolism are all fundamentally patterned after microbe-derived genetic platforms that evolved billions of years ago.

Aided by an infusion of funds by the Department of Energy, the human genome sequencing project helped to spawn huge investments in machines that could automate the sequencing of DNA. Indeed, with my colleagues at Rutgers University, we routinely sequence genomes at costs that are unimaginably low. When Sanger first began sequencing DNA, the cost was about 75 cents per nucleotide; by 2014 it cost less than 0.001 cent. To put this another way, in 2002, when the Human Genome Project was well underway, it was estimated that the cost of sequencing a human genome would be $100 million; it is now closing in on $1000 and almost certainly will be even lower in coming years.

Concomitant with the incredibly shrinking costs of sequencing, there was a huge increase in computing capacity and the interconnectivity of computers. Sequences of DNA can now be sent across the Internet in real time so that a best match can be assigned with a previously sequenced DNA molecule in milliseconds, and the newly identified sequence can be assigned a probable function within the cell.

With increasing computational capacity came more efficient and cheaper sequencing technologies and new algorithms for searching for genes. Indeed, the technologies became so cheap and the machines so widespread that there was surplus capacity in the national laboratories in the United States. That excess sequencing capacity soon spread rapidly across the world—to France, Germany, Great Britain, China, Japan, Korea, and India. What to do with it?

Shortly after the Human Genome Project began to get off the ground, David Galas, the head of the program in at the Department of Energy in Washington, D.C., came to Brookhaven National Laboratory to learn what the biologists were doing. The director of the lab asked me to give a short presentation about my effort to understand how a particular unicellular alga produces more or fewer specific proteins in response to changes in light, a phenomenon that is extremely important for phytoplankton in the oceans. Galas asked if I would convene a meeting to explore how the new sequencing and computational technologies could be applied to understanding the distribution of microbes in the environment. I gladly accepted the opportunity.

At the meeting, together with approximately sixty colleagues from across the country, I drafted a white paper that would ultimately lead to massive DNA sequencing of the microbes in the ocean, in soils, in the air, in lakes, on rocks, in ice—in virtually every environment conceivable. The result is that sequences from microbial genomes of the oceans are being produced at an unimaginable rate and tens of millions of new genes have been identified. This information is effectively a treasure trove of untapped biological potential that can be mobilized by humans to genetically engineer microbes to perform whatever task we so desire.

Literally, with a click of an electronic device, the sequence of a gene or many genes—indeed an entire genome—can be sent across a worldwide space to be analyzed, reshaped, and redistributed. Almost any gene can be synthesized and inserted into microbes. This free trade in gene function knows no bounds and has led to an escalating war with microbes.

As the sequencing of genes and genomes became so cheap and efficient by the beginning of the twenty-first century, scientists turned from sequencing the genomes of single organisms to sequencing those of natural microbial communities in virtually any place that was of potential interest. The list of genes detected by computer algorithms soared. Tens of millions of microbial genes have been identified on the planet, and the rate of their discovery is not yet showing signs of slowing down. This gene inventory represents a "parts list," the recipes for making any protein that nature has designed and is present in extant organisms. But could we make new parts? Ones that don't exist in nature and never have?

The short answer is yes.

A subset of the field of biology has morphed into something that seeks to design microbes, metabolism, and pathways within microbes—to make them more efficient or to give them new traits that never existed before. Could we make an organism that can degrade plastics? Or immobilize radioactive materials in a soil? Or create an alternative fuel? Or a new type of material? These issues are not academic. They are happening.

Thousands of laboratories around the world use variations of Lederberg's plasmids and Paul Berg's recombinant DNA to insert one or more genes into a microbe. The vast majority of these experiments are benign and are done to test hypotheses about how specific genes function. But a significant fraction of the horizontal gene transfers is carried out to manipulate a specific reaction in nature that we wish to change; for example, making a new photosynthetic organism from scratch.

The human genome sequence has revealed that we have virtually no unique genes. If we are lost, the world of microbes will proceed to carry out their functions and come to new steady states, whereby the ensemble of their metabolism maintains a habitable planet. Indeed, from an evolutionary perspective, human evolution is a temporary perturbation of biologically mediated chemical reactions. In short—we are freaks of nature that have disrupted natural geochemical cycles. Nevertheless, we need microbes.

We have become tinkerers of microbial evolution—and we don't understand what we are doing. The attempts are still academic exercises, but they are not trivial. For example, J. Craig Venter and his colleagues have worked on creating a microbe in which the genetic information is completely designed by humans with a computer, synthesized in a laboratory, and injected into a host cell that was genetically engineered to destroy its own genetic information. The host cell simply became a container for a totally human-designed genome.

Most synthetic biologists do not concern themselves with Earth systems—they are focused on making a better nitrogen-fixing microbe or, better yet, inserting the genes for nitrogen fixation directly into the cereal grains we depend on for our food. Synthetic biologists want to make a Rubisco that distinguishes between carbon dioxide and oxygen and spread the new and "better" Rubisco across the plant world. The list of alterations of microbes and other organisms attempted every day is virtually endless. Most of these efforts are noble attempts to develop a future that is sustainable for humans, but the lack of understanding of the unintended consequences for the evolutionary trajectory of life on Earth is very seldom considered.

Humans are a transient animal on this planet, and in our short history we have become one of the most disruptive biological forces since cyanobacteria began to make oxygen as a waste product of their metabolism. We are modern biological Bolsheviks. Like cyanobacteria, we can potentially unleash a Pandora's box of unintended consequences. I submit that rather than tinker with organisms that we can't reverse engineer, a much better use of our intellectual abilities and technological capabilities would be to better understand how the core nanomachines evolved and how these machines spread across the planet to become the engines of life.

Why? The microbes are the stewards of this planet, and we barely understand how they evolved a system of moving electrons and elements across its surface. Ultimately, that electron flow made Earth habitable for us. We have minimal knowledge of how that electronic circuit works, let alone know how to control it, yet in our hubris and insatiable need for more resources, we tinker with and inadvertently disrupt the circuit. Thankfully, there is so much redundancy built into the microbe-controlled electronic circuit that it is virtually impossible for us to seriously disrupt it, but that doesn't stop us from trying.

In the course of their evolutionary history, microbes have made this planet habitable for themselves and, ultimately, us. We are only passengers on the journey; however, we are tinkering with the organisms at the controls. It is only a matter of time that, without restraining ourselves, we will inadvertently design and release microbes that can fundamentally disrupt the balance of electrons in the global market. That would potentially be disastrous.

CHAPTER 11

Microbes on Mars and Butterflies on Venus?

There are not many questions in science that are as profound as, "Are we alone?"

The answer to that question would potentially forever change our view of ourselves. If we are not alone, what other life forms exist? How did they originate? What are the conditions on the planet where they live? As we try to understand how life originated on this planet and how the various emergent nanomachines became embedded into every organism that ever existed and continues to live on Earth, we are also asking, Did similar nanomachines evolve on other planets in our solar system or in planets orbiting stars far out in space? If so, how would we possibly know?

From the time Galileo discovered that Jupiter's moons orbit that planet and that the Earth is not the center of the universe, we have come a very long way in viewing our planet as a virtually insignificant dot of life in a sea of celestial haze. It is almost impossible to fully grasp the orders of magnitude required to reach the edges of the light emitted by stars born in an explosion at a single point about 14 billion years ago. Although our telescopes have become incredibly sophisticated instruments that peer into a vast space, the resolution of planets a few light years from us is still far less now than that of our best microscopes at the beginning of the twenty-first century. We can see objects move and estimate their size, but we can't yet tell if there is life outside of Earth. We really cannot yet tell if we are alone.

From scientific evidence that very few people truly understand, if at all, we now accept that the universe is expanding and that it contains billions of galaxies. Yet as far as we can tell, our planet is, for the moment, unique. It is the only planet we know that harbors life. Every bit of that life is based on nanomachines in microbes that churn out gases that are clear indicators of life. This planet is not only habitable, it is inhabited.

The question of Earth's singularity has haunted me for most of my life, as it does many of us. It is a question many children across the world ask as they look at stars and wonder how life began on this planet. It is a question that potentially can be answered, and the answers clearly lie in the evolution of microbes and their nanomachines, which have created a global electron market that has altered the composition of this planet's atmosphere, and thus the planet itself.

In our own solar system there are two neighboring planets that we can reach in a reasonable period of time with rocket-based landers: Venus and Mars. Today these two planets are very different from Earth, but that probably was not the case about three billion years ago.

Although the mass of Venus is a little more than 80% that of Earth, it has no liquid water on its surface. Venus is presently blanketed with an extremely thick layer of carbon dioxide, which is spewed from thousands of volcanoes. The layer of the gas is so thick that at the surface of Venus the pressure is approximately 100 times that at the surface of Earth. If we could stand on the surface of Venus, we would experience a pressure that is approximately equal to that 1000 meters beneath Earth's oceans. We would be squashed to about one-tenth our size. But we also would be boiled.

Being a greenhouse gas, the thick layer of carbon dioxide absorbs and traps solar radiation, making Venus the hottest planet in our solar system. It is so hot that lead would melt on its surface. But there is evidence that early in its history Venus was much cooler and possibly had liquid water on its surface. Whether it ever harbored life is an open question, but because of the extreme heat on its surface at present and alterations in its rocky surface, it is very unlikely that an unmanned lander could find evidence of any life that ever existed there. Mars, however, is a different story.

Today Mars is very cold and dry and has a very thin atmosphere. But it is also a much smaller planet than Earth, and its radioactive core has run out of fuel to make the interior of the planet hot enough to spew out carbon dioxide and other gases that are so critical for life. There hasn't been significant volcanic activity on Mars for more than 500 million years. Its surface is covered with lava from earlier volcanic eruptions and with loose grains of sand and dust; it is also dotted with boulders and craters. Mars has been a prime target for studies of life outside of Earth for several decades. Conceptually, life could have evolved on Mars and Venus as well as on Earth—but it appears that Earth alone won the lottery.

While we may be control freaks, we also are insecure and want to make sure that if we destroy this planet, we can find a home on a neighboring planet. Mars seems like a plausible candidate.

In 1975, six years after humans walked on the moon for the first time in history, NASA launched two satellites to Mars within a period of three weeks. The two satellites, Viking 1 and 2, were the most ambitious undertaking of the space program at the time. Each satellite was composed of two units: an orbiter and a lander. Over the next four years, the orbiters took more than 50,000 photographs of Mars and mapped the planet's surface. The landers weren't just demonstration satellites; they were equipped with instruments designed to find signatures of life on the red planet, whether it exists now or did in the past. These instruments were designed specifically to search for evidence of microbial life by tracking in Martian soil the gases they potentially produce as well as to see what types of organic matter they could have metabolized or produced.

The biological aspects of the program were extremely ambitious. The project was led by Gerald (Jerry) Sofen, a Princeton-trained biologist. During the Second World War, as an unarmed ambulance driver for the American army, Jerry, speaking Yiddish with a Cleveland accent, convinced a platoon of German soldiers to surrender in order to avoid being killed by the advancing Soviet Army. With that under his belt, it was easy to convince the NASA administrator to try to prove that life exists or existed outside of Earth.

At the time, the Viking mission to Mars cost over one billion dollars. Jerry assembled a science advisory board that included Joshua Lederberg and Harold Urey. Furthermore, Jerry had the vision to ask engineers to make instruments that could operate under the extreme conditions of Mars, and to make sure that the machines would be light enough to be launched into space and rugged enough to withstand years of high doses of radiation. These stringent conditions were not simple to meet.

Regardless, the instruments operated perfectly and sampled Martian soil for signs of organic matter, which would be the first indication of life. The initial results were tantalizingly promising, but after deeper consideration, it became apparent that the surface of Mars did not have any clear signs of life now or from the past. What it did show was evidence of liquid water and volcanic activity—two ingredients that almost certainly helped shape life on Earth. For the next several decades, a NASA mantra became "Follow the water." We have been following that mantra ever since. There have been several follow-up missions to Mars, but none of them, thus far, has found compelling evidence of life.

The Viking team realized that there was at least one potential, and potentially avoidable, problem with finding evidence of life on Mars. That problem is contamination from our own planet. Invariably, some microbe or other will hitch a ride on a satellite. NASA set out to make sure that this would never be an issue in the search for life with instruments that land on a planet. Indeed, the Viking landers were sterilized and meticulous care was taken to ensure that if there was evidence of life on Mars, we were not simply recording the activity of any microbial hitchhikers from Earth. But the problem was even more important if we wanted to return a sample from Mars to study it here.

On the third floor in NASA headquarters, at 300 E Street SW in Washington, D.C., there is an office with the presumptive but engaging title of Office of Planetary Protection. NASA's planetary protection officer (PPO) is charged with making sure that we minimize microbial contamination of our landers on Mars and other planets, moons, and former planets and their brethren. The PPO is also charged with making sure that if we bring samples back from those celestial bodies, they don't kill us or alter our planet forever. It's an interesting job, and I'm sure it makes for great opening lines at cocktail parties, but the job is serious, and for good reasons.

If we were to find evidence of life on Mars, would we also expect to find that evolutionary processes came to converge on exactly the same architecture of nanomachines? That would be very, very improbable, unless our ancestors originated on Mars and were transported to Earth on a meteorite, or vice versa. That may sound a bit far-fetched, but Martian meteorites are found on Earth. One of the most famous was discovered in 1984 in Antarctica by a group of geologists cruising the Allan Hills region of the continent on snowmobiles. It took a while to appreciate that the four-pound rock was not a typical meteorite.

The Allan Hills meteorite, denoted ALH840001, originated from rocks on Mars that were formed about 4.1 billion years ago. The meteorite was kicked out of Mars's gravitational field via an impact from a meteorite and landed on Earth about 13,000 years ago. It took about 10 years to understand the potential importance of the rock. In 1996, David McKay and his colleagues at NASA's Johnson Space Flight Center, near Houston Texas, suggested, based on microscopic analysis of the meteorite, that there was evidence of life on Mars.

What was the evidence? There were several lines. First, there are microscopic globules of carbonate in the meteorite. Carbonate formation on Earth requires liquid water. At the time, it was rather striking that water would have been present on early Mars, but even more striking was that inside the carbonate globules there were very small, wormlike structures that resemble fossil microbes. That was certainly surprising, but the structures are so small that it is difficult to understand how they actually could represent fossils of microbes. No known microbe on Earth is as small as the structures in the meteorite, and simple calculations suggest that if such cells actually existed, their genome would have been incredibly streamlined. However, a third line of evidence was based on the presence of very small grains of magnetite, an oxide of iron that is commonly found in geological settings. The shapes of those grains are so precise that they resemble those produced by magnetotactic bacteria. Moreover, the bacteria that produce magnetite form small chains of the crystals within the cells, resembling a microscopic chain of pearls. The chains of magnetite allow the cells to sense magnetic fields. Some of the crystals of magnetite in the meteorite appear to be arrayed in chains very much like those found in magnetotactic bacteria and are arguably the strongest evidence of life.

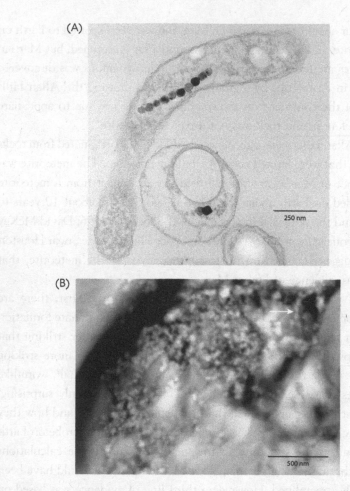

FIGURE 38. (A) An electron micrograph of a series of magnetic (magnetite) particles lined up in a bacterium to form a *magnetosome*—a structure that allows the cell to sense magnetic fields. The structure is extremely small, precise, and highly organized. It is made and controlled by the bacterium. (Courtesy of Atsuko Kobayashi) (B) A scanning electron micrograph of polished samples from the Allen Hills meteorite (ALH84001) reveals a chain of elongated magnetite particles in the upper right-hand corner aligned along an axis (arrow). This structure is similar to those found in magnetotactic bacteria. (Courtesy of J. Wierzchos and C. Ascasco)

The paper describing the potential evidence of life on Mars, published on August 6, 1996, in *Science* magazine, one of the most reputable scientific journals in the world, certainly got people's attention and sparked a huge, renewed interest in the search for life on the red planet. The president of the United States at the time, Bill Clinton, held a press conference on the south lawn of the White House a day after the publication of the paper and said, "Today, rock 84001 speaks to us across all those billions of years and millions of miles. It speaks of the possibility of life. If this discovery is confirmed, it will surely be one of the most stunning insights into our universe that science has ever uncovered. Its implications are as far-reaching and awe-inspiring as can be imagined. Even as it promises answers to some of our oldest questions, it poses still others even more fundamental." It hit the front pages of every major newspaper around the world and injected a renewed sense of purpose for NASA.

Although the interpretation of the microscopic structures in the Allan Hills meteorite remains extremely controversial, it focused a lot of attention on two of the core questions of science: "Where did life originate?" and, "Are we alone?" Many scientists also asked, "Are we Martians?" Joe Kirschvink sometimes argues that all life on Earth descends from the contamination of our planet by a Martian meteorite.

Follow-up analyses of ALH84001 have been difficult to reconcile with life as we know it. Most geologists now dismiss that the meteorite has compelling evidence of fossil microbes, but the process that led to the formation of the precisely formed chains of magnetite remains enigmatic. Regardless, the discovery of that meteorite certainly sparked a renewed search for potential past and present life on Mars.

Jerry Sofen convinced the NASA administrator, Dan Goldin, to send new landers to Mars and to develop a search for life elsewhere in the universe. But just to make sure this wouldn't be a passing interest for NASA, Jerry convinced NASA to develop a program in *astrobiology*, and in 1998 he oversaw the creation of NASA's Astrobiology Institute. One of the most interesting and challenging roles of the institute is to search for evidence of life within our solar system and beyond.

In the first two decades of this new millennium, NASA successfully landed several new mobile rovers on Mars, each equipped with increasingly sophisticated instruments designed to search for evidence of life. There has been a lot of effort to find gases, such as methane or nitrous oxide, which are indicative, but not proof, of microbial life. Thus far, the signs have not been positive, let alone conclusive. These missions will continue for the next decades, and plans are being made to bring samples of Martian soil and rocks back to Earth for more extensive analyses. These missions are engineering feats, and we have learned a lot about Martian history. But in the meantime, we also set our sights further afield in the quest to answer the question, "Are we alone?"

In 1972, as part of the Apollo Mission series, NASA launched the first space-based telescope. The instrument recorded ultraviolet light, which is blocked on Earth's surface because our atmosphere absorbs much of the light in that portion of the spectrum. It was the beginning of one of the most important series of discoveries of our universe since Galileo's first descriptions of the Jovian moons.

Telescopes are designed to detect light, but by not having to peer through Earth's atmosphere, space-based telescopes have remarkable resolution of very distant objects. They can detect extremely small differences in light from stars in our galaxy, the Milky Way.

In 1988, three Canadian astronomers, Bruce Campbell, Gordon Walker, and Stephenson Yang, reported a periodic change in the wavelengths of light recorded from a binary star, Gamma Cephei, approximately 45 light years from Earth. Binary stars contain two stars that orbit around a center of mass; they are very common. The change in the wavelengths that the three astronomers detected was a result of the light being received slightly faster and slower: a Doppler shift. They suggested that the Doppler shift was a result of a planet orbiting one of the stars, and in so doing forcing the star to undergo an orbit of its own. They named the planet Gamma Cephei Ab. The report was met with some skepticism, and it was not until 2002 that it was confirmed. Gamma Cephei Ab was the first planet discovered outside of our solar system, but by 2014 there were approximately 2000 confirmed reports of extrasolar planets, with hundreds being more detected each year. But how would we know if a planet harbors life? They are so far away that we couldn't possibly land rovers on even the closest extrasolar planet in our children's, children's, children's lifetimes. Let's see why.

Two satellites, Voyager 1 and 2, which were launched in 1977, are just leaving our solar system after traveling approximately 18 billion kilometers. That's an average velocity of about 500 million kilometers per year, or about 35,000 miles per hour. At that speed, they could reach the star closest to Earth, Proxima Centuri, which is 4.2 light years away, in about 80,000 years. I don't think we can wait that long to find out if we are alone, especially if Proxima Centuri doesn't have planets that harbor life. Fortunately, astronomers have alternative approaches to searching for life outside of our solar system.

One, just discussed, is the Doppler shift from stars due to a change in the star's orbit resulting from a neighboring body orbiting the star. The effect is rather straightforward: a star with a planet orbiting around it also has an orbit. The orbit of the planet can be detected from the changes in the wavelengths of light arising from the spectral lines in the star. As the star moves ever so slightly toward us (or our space-based telescope), the spectral lines are shifted toward the blue (shorter wavelengths). As it moves away, the spectral lines are shifted toward the red (longer wavelengths). The larger the planet, the larger the effect, and hence, most of the planets identified to date are massive, about the size of Jupiter or Saturn. Those planets are hundreds of times more massive than Earth, and most of them don't have any land or oceans; they are gaseous. It's hard to imagine life existing on such planets.

However, there is a second method to detect planets. It is based on the very tiny amount of light that is blocked as a planet crosses in front of its star. As difficult as it is to believe, both space-based and Earth-based telescopes can detect this so-called transit, even from stars tens of light years away, which, from an astronomical perspective, is right in our backyard. The principle of the measurement is relatively simple—when a planet crosses in front of its star, the light from the star is slightly lower than when the planet is on the other side of the star. The difference between the amount of light detected with and without the planet between our telescopes and the planet's star provides a basis for determining the size of the planet. The larger the planet, the more light is blocked. If one knows the size of the planet from its transit and the mass of the planet from the Doppler shift due to the orbital velocity, then the ratio of the two, mass per unit size, gives a clue as to the planet's density.

Dense planets are rocky planets, like ours, and rocky planets can potentially harbor life. But there are a few more characteristics we can reasonably infer from telescope observations. One of the most important is the transit time of the parent around its star. Earth, which is the third planet from the Sun, has an orbit time of 365.26 Earth solar days. Venus has an orbit of 224.7 days, while Mars orbits the Sun every 697 Earth solar days. In fact, if one examines the orbital times for all the planets in our solar system, the time of orbit is directly related to the planet's distance from the Sun, regardless of the planet's mass. The longest orbit time is Neptune's (since Pluto is no longer considered a planet), which at 60,200 Earth days, corresponds to about once every 164 Earth years. In other words, a human will not witness one complete orbit of Neptune in his lifetime. However, if the transit time of a planet is related to its distance from the star, then we can determine how much solar radiation the planet could potentially intercept. And that's a big deal.

Our two nearest neighbors, Venus and Mars, don't have liquid water on their surfaces any more. One is too hot, the other too cold. In this Goldilocks world of a perfect planet, Earth has maintained a relatively constant temperature that has allowed liquid water to remain on its surface for as long as we can tell. One reason is because we are not too close to our star; another is that the greenhouse gases in our atmosphere have adjusted over time. And that in itself is remarkable.

The concentration of greenhouse gases, especially carbon dioxide and methane, must have been much higher three billion years ago, when the Sun was less luminous. On Venus, the carbon dioxide concentration continued to rise as volcanoes spewed the gas into its atmosphere. This caused water to evaporate, and at the top of the atmosphere the water was split by ultraviolet light from the Sun to form hydrogen and oxygen. Hydrogen, being the lightest element, can escape the planet's gravitational field and be swept out into space. The oxygen would react with the rocks on the planet's surface. Over time, this process would boil away the oceans of Venus. Such a phenomenon will almost certainly happen on this planet in a few billion years as our star slowly heats up and increases in luminosity. But our planet has been habitable for more than four billion years, whereas Mars and Venus no longer have liquid water on their surfaces.

One of the reasons liquid water has been maintained on Earth for so long is a consequence of the feedback between the evolution of microbes and the atmosphere of Earth. As microbes increasingly developed a global electron market, the gas composition of the atmosphere changed. Carbon dioxide was removed from the atmosphere, and about 20% of it was converted to organic matter and buried in rocks. Oxygen, which is not a greenhouse gas, accumulated. These changes allowed for the emergence of animal life on Earth.

Although we can be fairly sure that there are no butterflies on Venus under the present conditions, and probably there never were, are there planets outside of our solar system that harbor life? If so, what would be the evidence?

If we could determine the composition of a planet's atmosphere, along with its mass and distance from its star, we could potentially infer whether there is life outside of our solar system. Amazingly, that appears to be doable. The primary method to detect planetary atmospheres takes advantage of the transit of the planet in front of the star which, from the perspective of the observer, is an eclipse. During the eclipse, light from the star shines across the thin film of the planet's atmosphere. Gases in the atmosphere absorb the light, and the difference between the spectrum of light from the star alone and that with the eclipse of the planet can be used to infer the gas composition of the planet's atmosphere. Using several sophisticated techniques, one can help reduce the light from the background star and very precisely determine the spectrum of the light detected by the telescope. These measurements require not only a significant investment in instrumentation, but also lots of precious observing time on telescopes. Consequently, we have far less information about the atmospheres on extrasolar planets than we do about the census of extrasolar planets themselves. We have been able to detect planetary atmospheres that contain water vapor, carbon monoxide, carbon dioxide, methane, and even acetylene. Most of these planets are gaseous ones very close to their parent star. They are very large and very hot. Thus far, none are in the habitable zone of their stars, and none are candidates for harboring life, but that almost certainly will change during the next decade or so, as we discover more planets and our observational tools become more sophisticated.

Evidence that life could exist on an extrasolar planet is whether the atmospheric gas composition is at equilibrium. Let me be clear, by "equilibrium," I mean that the gases can be readily produced by the planet's own geological setting. For example, on Earth, volcanoes emit carbon dioxide and methane, and the heat from the Sun vaporizes liquid water whether or not there is life. These gases are not by themselves indicative of life. However, the alteration of our atmosphere by microbes, long before there were plants and animals, gives us some idea about what gases to search for on extrasolar planets in the habitable zone, a place where a planet is close enough to its star to allow liquid water to exist on the surface.

One of the obvious is the presence of molecular oxygen, which has led to the production of stratospheric ozone on Earth. Detection of ozone on a terrestrial planet within the habitable zone would be hard to accommodate without assuming the presence of life. Ozone is not a gas that would be maintained by any mechanism that we understand under equilibrium conditions. Another candidate that is not at equilibrium is nitrous oxide. If a terrestrial planet containing both nitrous oxide and methane in its atmosphere were to be detected, it almost certainly could harbor life.

In January 1613, four years after Galileo discovered that the moons of Jupiter orbited that planet, he observed a planet in our solar system that could not be seen with the naked eye. That planet, Neptune, is about 4.5 billion kilometers from Earth and, like it, orbits the Sun. Four hundred years later, astronomers estimate that there are about 144 billion planets in the Milky Way alone. While that number is staggering, there are more than 100 billion galaxies in the known universe. That makes the odds that we are alone very small indeed. If we are the only planet harboring life, Earth hit the lottery of life in more than 10^{22} chances. I would bet that there are other winners nearby in our own galaxy—but I don't bet.

Given the odds, the discovery of gases far from equilibrium on a terrestrial planet in the habitable zone is almost inevitable. That discovery will be transformative for us as humans. It will make us think about what makes this planet so rare, and yet, perhaps not so rare. But it will also force us to understand that life can evolve independently many times in many places. We will know that some nanomachines evolved elsewhere to move electrons across a planetary surface and, in so doing, altered the gas composition of a planet. While we can never be absolutely sure, we can speculate that a microbial system was probably responsible for making that planet conducive for life, and perhaps even for complex life.

Our phylogenetic trees of life are confined to this planet. It is beyond belief that we share a common origin with life on planets many light years away. And if that is so, are there several possible solutions to the origins of life?

Life, unleashed, must find a way to exist on another planet. But how?

As long as the basic systems work, that planet will harbor some reactions that will persist, independently, from all other life in the celestial haze. The systems include some geological recycling of materials for organisms. On Earth, that process is tectonics. It does not mean that is the only process, but it is the only process we know of that works on time scales of billions of years. It must also include an atmosphere—or some fluid that acts like a wire to connect organisms' metabolisms across the planetary surface.

Life on Earth is both fragile and resilient. I know, for sure, that this planet harbors butterflies, and those, apparently fragile, organisms, have been here for more than 200 million years. But like us, they depend upon microbial machines for their very existence. Thanks be to microbes for making this speck of detritus in the stardust of the universe a great place to live for their overgrown relatives, the animals and plants that temporarily decorate and rent the small dot from their microbial ancestors, who maintain it for their future relatives.

The connected contingencies for life will almost surely not be found during an elevator ride in a New York City housing project. But such contingencies do allow us to explore the world in which we live and search for life outside our planet, in the sources of light from distant stars and their planets. Whether we will find "intelligent" life is another question. Intelligent life is probably a very rare commodity in our galactic neighborhood. It has evolved on Earth only in the last couple of million years, and it is only within the past century that we have developed technologies that have transformed the planet forever.

If we are alone, we need to understand our inadequacies. If we are not alone, we need to be humbler. Regardless, as one eukaryote talking to fellow eukaryotes, we are all macroscopic bodies, and our existence is made possible only by the evolution of microscopic nanomachines that evolved a long, long time ago, in microbes. They are our true ancestors and the true stewards of life on Earth.

Further Readings

|||||||||

CHAPTER 1

The 1785 Abstract of James Hutton's Theory of the Earth. C. Y. Craig, editor. 1997. Edinburgh University Press. This thirty-page essay was transformational and inspired Lyell.

Darwin and the Beagle. Alan Moorhead. 1983. Crescent Press. A lively and very readable narrative of the life of Darwin aboard the *Beagle*, which is, in some ways, better than Darwin's original account.

Measuring Eternity: The Search for the Beginning of Time. Martin Gorst. 2002. Broadway Publisher. A well-written and well-researched book about how we came to know the age of the Earth.

On the Origins of Species. Charles Darwin. 1964. Harvard University Press. A facsimile of the first edition.

Principles of Geology. Charles Lyell. 1990. University of Chicago Press. This is a reproduction of the original set of books by Lyell, with the illustrations by the author. It is a bit of a tedious read.

Seashell on a Mountaintop: How Nicolas Steno Solved an Ancient Mystery and Created a Science of the Earth. Alan Cutler. 2004. Plume Press. A great historical account of how fossils came to be discovered.

CHAPTER 2

"The discovery of microorganisms by Robert Hooke and Antoni van Leeuwenhoek, fellows of the Royal Society." H. Gest. *Notes Rec. R. Soc. Lond.* (2004) 58: 187–201. doi: 10.1098/rsnr.2004.0055. A wonderful account of Leeuwenhoek and his friendship with Hooke.

Microbe Hunters. Paul de Kruif. 1926. Harvest Press. The classic introduction to some of the early pioneers of microbiology especially in relation to diseases. It is a bit dated now.

Micrographia—Some Physiological Descriptions of Minute Bodies Made by Magnifying Glasses with Observations and Inquiries Thereupon. Robert Hooke. 1665. Reprinted 2010. Qontro Classic Books. A reproduction is available for free online: www.gutenberg.org/ebooks/15491.

CHAPTER 3

The Age of Everything: How Science Explores the Past. Mathew Hedman. 2007. University of Chicago Press. A great book that explains how we know the age of civilization, the age of the Earth, and the age of the universe.

Darwin's Lost World: The Hidden History of Animal Life. Martin Brasier. 2010. Oxford University Press. A very accessible and personal account of the evolution of animals.

Life on a Young Planet: The First Three Billion Years of Evolution on Earth. Andrew Knoll. 2004. Princeton University Press. A delightful account of the Precambrian world of microbes.

CHAPTER 4

Aquatic Photosynthesis. P. G. Falkowski and J. A. Raven. 2007. Princeton University Press. A textbook that describes the fundamentals of photosynthesis from both mechanistic and evolutionary perspectives. Not for the faint-hearted.

Life's Ratchet: How Molecular Machines Extract Order from Chaos. Peter M. Hoffmann. 2012. Basic Books. A very accessible book explaining the action of molecular machines.

"There's plenty of room at the bottom: An invitation to enter a new field of physics." R. P. Feynman. 1960. Available online at http://www.zyvex.com/nanotech/feynman .html. A wonderful essay on nanomachines.

What Is Life? The Physical Aspect of the Living Cell. Edwin Schrodinger. 1944. Cambridge University Press. Available online at http://whatislife.stanford.edu/LoCo_files /What-is-Life.pdf. This is the classic book in which a theoretical physicist attempts to understand how life works.

CHAPTER 5

Cradle of Life: The Discovery of Earth's Earliest Fossils. J. William Schopf. 1999. Cambridge University Press. A personal account of how we discovered Precambrian fossils of microbes.

Eating the Sun: How Plants Power the Planet. Oliver Morton. 2007. HarperCollins. An excellent account about the photosynthetic processes and how they transformed Earth.

Oxygen: A Four Billion Year History. D. E. Canfield. 2014. Princeton University Press. A delightful book exploring how oxygen became such an abundant gas on Earth.

Oxygen, The Molecule That Made the World. Nick Lane. 2002. Oxford University Press.

CHAPTER 7

Microcosmos: Four Billion Years of Microbial Evolution. Lynn Margulis and Dorian Sagan. 1997. University of California Press. An explanation of the importance of microbial evolution and symbioses.

CHAPTER 8

Lives of a Cell: Notes of a Biology Watcher. Lewis Thomas. 1978. Penguin Press. A classic collection of essays by Thomas that is charming, witty, and inspirational.

Power, Sex, Suicide: Mitochondria and the Meaning of Life. Nick Lane. 2005. Oxford University Press. A great book describing how mitochondria work and their role in shaping the lives of eukaryotes.

Wonderful Life: The Burgess Shale and the Nature of History. Stephen J. Gould. 1989. W. W. Norton. One of Gould's best books and a fascinating glimpse into the history of paleontology.

CHAPTER 9

The Alchemy of Air: A Jewish Genius, a Doomed Tycoon, and the Scientific Discovery That Fed the World but Fueled the Rise of Hitler. Thomas Hager. 2008. Three Rivers Press. A history of Haber and Bosch and the chemical reaction that led to the commercial production of nitrogen fertilizer.

From Hand to Mouth: The Origins of Human Language. Michael C. Corballis. 2003. Princeton University Press.

The Genesis of Germs: The Origin of Diseases and the Coming Plagues. Alan L. Gillen. 2007. Master Books. A book describing how microbial diseases evolve and spread. Not recommended as a bedtime reader.

Microbes and Society. Benjamin Weeks. 2012. Jones and Bartlett Learning.

CHAPTER 10

The Double Helix: A Personal Account of the Discovery of the Structure of DNA. James D. Watson. 1976. Scribner Classics. The title says it all.

Introduction to Systems Biology: Design Principles of Biological Circuits. Uri Alon. 2006. Chapman and Hall/CRC Press. A rather dry book; not for the faint hearted.

Life at the Speed of Light. J. Craig Venter. 2013. Viking. A historical and personal overview of how synthetic biology became embedded in the culture of contemporary science.

Regenesis: How Synthetic Biology Will Reinvent Nature and Ourselves. George Church and Edward Regis. 2014. Basic Books. A historical overview of geology, as captured by a chemist, and the ability of humans to genetically transform microbes and ourselves.

Rosalind Franklin and DNA. Anne Sayre. 1975. W. W. Norton. The back story about how the structure of DNA was discovered.

CHAPTER 11

How to Find a Habitable Planet. James Kasting. 2010. Princeton University Press. A great read on the logic of how astronomers can identify life outside of our solar system.

The Life of Super-Earths: How the Hunt for Alien Worlds and Artificial Cells Will Revolutionize Life on Our Planet. Dimitar Sasselov. 2012. Basic Books. Sasselov is an astronomer; this is his account of the origins of life and how it might be found elsewhere in our galaxy.

Rare Earth: Why Complex Life Is Uncommon in the Universe. Peter Ward and Donald Brownlee. 2000. Copernicus Books. The authors present a pessimistic view of the number of planets that harbor complex life.

Index

‖‖‖‖‖‖‖‖

Page numbers followed by "f" and "t" indicate figures and tables.